WORLD ETHICS AND CLIMATE CHANGE

EDINBURGH STUDIES IN WORLD ETHICS

WORLD ETHICS AND CLIMATE CHANGE

FROM INTERNATIONAL TO GLOBAL JUSTICE

Paul G. Harris

EDINBURGH UNIVERSITY PRESS

© Paul G. Harris, 2010

Edinburgh University Press Ltd
22 George Square, Edinburgh

www.euppublishing.com

Typeset in Times by
Iolaire Typesetting, Newtonmore, and
printed and bound in Great Britain by
CPI Antony Rowe, Chippenham and Eastbourne

A CIP record for this book is available
from the British Library

ISBN 978 0 7486 3911 3 (hardback)
ISBN 978 0 7486 3910 6 (paperback)

The right of Paul G. Harris
to be identified as author of this work
has been asserted in accordance with
the Copyright, Designs and Patents Act 1988.

CONTENTS

PREFACE

More than two decades of international negotiations have failed to stem emissions of greenhouse gases that are causing global warming and climate change. This book identifies a way to escape this ongoing tragedy of the atmospheric commons. In it I attempt to take a different approach to the ethics and practice of international environmental justice. I propose significant adjustments to the existing international climate change regime, in the process drawing support from cosmopolitan ethics and global conceptions of justice. Ultimately, the book is an argument for a new, cosmopolitan kind of diplomacy and politics that sees people, rather than states alone, as the causes of climate change and the bearers of related rights, responsibilities and obligations.

As the title suggests, this book is about *world ethics* – an exploration of moral values, norms and responsibilities that apply globally (cf. Dower 2000: 265). It is about a particular kind of world ethic – namely, cosmopolitanism – which asserts that human beings ought to be at the centre of moral calculations, that ethical obligations and responsibilities are not defined or delineated by national borders, and that universal values exist and (for many cosmopolitans) ought to guide the behaviour of people and states (cf. Dower 1997: 561). Importantly, my exploration of world ethics is not a work of abstract philosophy. The aim here is to apply concepts of right and wrong devised by others to the problem of climate change. I take a position and spend some time defending it using selected ethical philosophy, but whether the position is right or wrong from a rigorous philosophical perspective is less important to me than whether it is right or wrong in most people's minds. As such, the book is about *practical* world ethics – what we ought (or ought not) to do as well as why we ought to do it (or not). More specifically, the book is about what cosmopolitanism says regarding the causes of global warming (does it matter who or what causes global warming?), the consequences of it (does it matter who is affected, and how they are affected?) and the

vii

politics of climate change (how should we distribute the associated burdens, and among whom?). The book is about using cosmopolitanism as a practical tool to help us to understand climate change and to craft better climate change policies. My objective is to deploy world ethics in the service of environmental protection and human welfare, not to use an environmental issue to explore world ethics (although this will happen inadvertently).

The book is also about international politics, in particular the manner in which cosmopolitan considerations ought to be and can be integrated into interstate discourse and policy. Here I am concerned with moral principles, notably what they tell us about the rights and obligations of persons everywhere regardless of nationality, as well as what those principles imply for political institutions of states and the international community. The message of the book is that cosmopolitanism offers people and governments a pathway by which to escape the tragedy of the atmospheric commons.

Insofar as this book is a critique of the status quo, it attempts to do two things. First, I seek to reveal the ineffectual overemphasis on the state and the immoral neglect of the person in the context of climate change policy making. Second, to the extent that persons are being given greater consideration in climate ethics and politics, I seek to show how the discourse – and theorising and policy formulation – overlook (to put it charitably) the role of some of the world's most affluent and polluting persons: the hundreds of millions of well-off people living outside the developed states. I shine a light on these people not because they are more important than well-off people in the developed countries but because they have a much greater practical significance than they did even a decade ago. We cannot continue to ignore them for practical reasons; we ought not do so for ethical ones.

Living in Asia has helped me to see more clearly that the number of affluent people in the developing world is now very large, and that it is growing rapidly. It is simply not practical – and not just – to let the most affluent *people* in poorer countries (which now include me, living very comfortably in China) avoid this issue simply because the affluent *states* have been recognised to be legally to blame for most present and much future climate change. We can, of course, say that the wealthy states are also *practically* to blame for most climate change, both by aggregate (historical) and average per-capita measures. However, this may be the wrong, or at least a very much inadequate, discourse. To talk of climate justice in that way frames the issue in terms of states, which is acceptable *only if it is supplemented* with much more talk of the obligations of affluent *individuals* and critiques of their consumption choices. In short,

I want to direct more attention to the obligations of affluent people *everywhere*, including those people who are in places where their obligations are almost totally ignored ethically, legally and even practically.

I come at this as a student of international affairs, and as someone who has dabbled in questions of international (that is, interstate) environmental justice and who is frustrated with the failure of international politics to address climate change and its consequences effectively. One impetus is my dissatisfaction with the usual international justice arguments (including my own!) that suggest solutions but often ignore the individual level of obligation and, especially, action. This book is, therefore, a critique of the status quo statism of most official and scholarly discourse, as well as national and international action, on climate change. I acknowledge that many of the solutions to climate change will have to involve states. However, this reality need not absolve capable *individuals* from explicit responsibility and obligation; nor should it prevent diplomats, activists and scholars, along with laypersons, from discussing it and attempting to implement it personally.

I can anticipate that a reviewer may say something like this: 'He's letting Americans off the hook and asking Chinese to stop polluting. How shocking!' In two respects this would be true (continuing to use the quite apt examples of Americans and Chinese): I do think that *very poor people* in the United States – but not the United States as a whole (as a nation-state) – ought to be let off the hook to a degree. After all, they do not contribute much to the problem or at least have no choice about doing so. To be sure, most people in China, who are very poor by any measure, deserve to *increase* their energy use and their material consumption – as do a small minority of Americans. But I also believe that *affluent and capable* people in China, of which there are now many millions, ought to reduce their pollution (and do more). The upshot of my argument is that nearly all Americans (and of course Australians, Britons, Canadians et al.) need to do much more to reduce their material consumption and pollution, but so should a rapidly growing minority of Chinese (and Brazilians, Indians, South Africans et al.). What is important is that everyone everywhere be treated equally, *ceteris paribus*.

I write this as a concerned resident of our planet. I invoke philosophers to find ethical support for expanding the locus of obligation to act on climate change and to aid those who suffer from it, even though I realise that it will be very difficult to foster action on that obligation. I do not undertake the task of presenting a fully developed philosophical exegesis. Mine is a normative argument with a practical objective, which is necessitated by the failure of existing international arrangements and the suffering (human and non-human) that is under way and will arise from that failure.

This book builds on *International Equity and Global Environmental Politics* (Harris 2001a), and many of the ideas were first exercised in other books and articles (cf. Harris 1996, 1997a, 1997b, 1999a, 1999b, 2000a, 2000b, 2001b, 2002a, 2002b, 2002c, 2003a, 2003b, 2004, 2006, 2007a, 2007b, 2008a, 2008b, 2008c, 2008d, 2009). I am grateful to all of the people who commented on that earlier work. I am indebted to Nicola Ramsey, senior commissioning editor at Edinburgh University Press, for adopting the project, the Edinburgh University Press Committee for endorsing it, and a number of interns and other helpful people at the press who have ushered the book through the production process. I am especially grateful to Nigel Dower, editor of the Edinburgh Studies in World Ethics series, for seeing merit in this project and for giving me very thoughtful advice on how to strengthen the manuscript. I am also grateful to anonymous reviewers commissioned by Edinburgh University Press. My thanks go to Jonathan Symons for comments on an earlier draft and Felix Tsang for help in gathering documents. As always, I am most thankful for the daily support of K. K. Chan, not least because research, writing and teaching, not to mention myriad administrative duties, often leave little time for anything (or anyone) else.

This book was written as a research monograph. To assist lecturers and students who will be reading it as part of academic courses and seminars, a companion *World Ethics and Climate Change Learning Guide* is freely downloadable from the Edinburgh University Press website at www.euppublishing.com.

Two central themes in this book are environmental sustainability and individual obligations of global justice. In an attempt to implement the former, the book is printed on paper from certified sustainable sources. To act on the latter, all of my royalties will be directly paid by Edinburgh University Press to Oxfam in support of their work among the world's poor, including those people most harmed by climate change. These are by no means acts of altruism, charity or generosity. They are acts of cosmopolitan obligation in a very fragile – and all-too-frequently unjust – world.

Paul G. Harris
Lantau Island
South China Sea

INTRODUCTION

The ecological underpinnings of the Earth are under monumental assault by human beings. As a consequence, the world is caught in a truly profound dilemma. Decades-long efforts by governments and the international community to cooperate in protecting the global environment have failed to bring about robust action to limit greenhouse gas pollution causing global warming and climate change. While pursuing apparently logical economic and social development, and by acting in ways that are assumed to promote the interests of states and their citizens, humanity continues dangerously to alter the Earth's atmospheric and climate systems, with profound consequences for human well-being and, for many millions of people, even survival. One reason for this tragedy of the atmospheric commons is the preoccupation of governments and societies with political independence and national sovereignty, the dominance of an international system premised on that sovereignty, and a failure adequately to recognise twenty-first-century realities, notably rapidly expanding numbers of new consumers in the developing countries that are adding greatly to the greenhouse gas pollution that has long come from people in the developed countries. The dilemma brought on by this preoccupation with states and their sovereign rights requires an alternative pathway leading to environmentally sustainable development that is agreeable to both rich and poor countries and to their peoples.[1]

As part of efforts to find this pathway, this book's project is to explore the role of justice in the world's responses to climate change, and in particular to introduce and explain an alternative strategy for tackling climate change that is more principled and practical than the prevailing doctrine, and that may be much more politically acceptable to governments and citizens than are existing responses to the problem. This alternative strategy is premised on cosmopolitanism. A cosmopolitan ethic, and its practical implementation in the form of global justice,

1

offers both governments and people a path to sustainability and successful mitigation of the adverse impacts of climate change.

The chapters that follow look at the problem of climate change through the prism of world ethics, which can be characterised as 'the exploration of the complex moral values, norms, and responsibilities that we acknowledge in regard to the relations between states and the relations individuals have with one another and the natural world on a global scale' (Dower 2000: 265). The chapters attempt to frame and answer a question posed by Brian Barry (2008: p. ix): 'If we accept the dominant view that each member of the human species has an equal share of the capacity of the earth to absorb carbon emissions, what ethical and policy proposals flow from the "inconvenient truth" that a minority of the world's population (mostly living in wealthy, industrialised countries) are not only using up a disproportionate (and therefore unfair) amount of this resource but are also the major cause and beneficiary of that unfairness?'[2] The particular ethic that follows from Barry's question, and the one that informs the book (particularly Part III) and its conclusions, is a global one with two aspects, namely, 'certain values and norms that are universal, in that they are applicable to all human beings everywhere', regardless of the states in which people live, and 'certain duties or responsibilities that are global in scope, in the sense that individuals, states, and other bodies have, in principle, duties towards all others in the world' (Dower 2000: 265).

The global ethic that permeates the book is a cosmopolitan one that assumes all of us to be world citizens in the sense that we are all 'members of one global society, with duties towards one another . . . National borders and identities, therefore, are not of ultimate moral significance' (Dower 2000: 265). I go beyond still important questions of *international* climate justice to explore cosmopolitan or *global* climate justice. I will try to do what Molly Cochran (1999: 21) says that cosmopolitans do: 'seek to interrogate and complicate the value conferred upon sovereign states in the contemporary international system, since cosmopolitans take individuals, not states, to be the starting point for moral consideration.' Thinking in cosmopolitan terms directs our attention to the many millions, and possibly billions, of people harmed by climate change and whose rights to life and well-being are violated by it. Cosmopolitanism also requires us more carefully and explicitly to consider the obligations of the world's affluent people – those who consume the most (usually things they do not need) and generate the most atmospheric pollution per capita – to do much more to address this problem, regardless of whether they live in affluent or poor states. Cosmopolitan justice can locate more obligation to act on climate change, and to aid those people who suffer from it, in affluent individuals *everywhere*.[3]

Part I of the book sets the stage by describing major practical and ethical challenges of climate change, looking in particular at the causes, impacts and injustices of climate change in the context of broader considerations of how justice does and should obtain in world affairs. Part II of the book is about climate change and *international* justice. It describes and critiques the interstate, communitarian doctrine underlying and guiding ongoing international negotiations and policy responses to climate change. Part III is about climate change and *global* justice. It explores an alternative, cosmopolitan perspective of the problem, in so doing critiquing the routine and increasingly anachronistic preoccupation (even of many cosmopolitans) with people in developed countries.

THE CHALLENGE

Chapter 1 briefly summarises the monumental problem of climate change, focusing on its impacts, particularly for the world's poorest and weakest countries, communities and people. It describes some of the causes and consequences of climate change and identifies some of the reasons why climate change is a matter of international justice – and injustice. Global warming is causing increasingly significant ongoing climate change that will become profoundly damaging to human well-being in this century and beyond. While all regions of the world will be impacted by climate change, it is the poorest regions and poorest people that will suffer the most. The world's wealthy countries and people will, in most cases, be able to adapt to climate change, or at least to survive it. In contrast, the poorest countries, the poorest regions within them, and the world's poorest individuals, most of them in Africa and developing parts of Asia and Latin America, will suffer and often die as a consequence of climate change. Importantly, those who will suffer the most from climate change – the world's poor – are the least responsible for it. Historically it has been the world's wealthy states and their citizens that have polluted the atmosphere, often as a result of conspicuous consumption and other activities that are not essential to life or happiness (and indeed often undermine them, as when people neglect family and friends to garner wealth and possessions or when they consume foods that are both bad for the environment and bad for their health). Now the burgeoning middle and wealthy classes of the developing world – the world's new consumers – are adding to this pollution, leading to explosive growth in greenhouse gas emissions.

The causes and consequences of climate change raise major practical challenges for societies and governments. They also raise the most profound questions of international and global justice yet encountered in human history. Chapter 2 frames these questions in the context of wider considerations of ethics and justice in world affairs. While communitarian conceptions of ethics and justice largely restrict the scope of our obligations to fellow citizens, cosmopolitan accounts of justice extend those obligations much farther, in the process substantially discounting or even rejecting the moral significance of the states in which people live. But these different accounts of how far the scope of justice should extend do not tell us very much about what is meant by justice, which is a concept subject to different, sometimes competing, definitions. Chapter 2 attempts to define justice sufficiently to understand how the concept is germane, ethically and practically, to climate change diplomacy and policies. In very general terms, justice in this context is about how the benefits and burdens associated with climate change are distributed among states, people and other actors. Drawing on several common accounts of justice (for example, utilitarianism, Kantianism, basic rights), the chapter shows how one can conclude, from a range of perspectives, that climate change is very much a matter of justice, and indeed of *in*justice.

INTERNATIONAL JUSTICE

Chapter 3 describes the concept of environmental justice and the interstate doctrine upon which it has been layered, as governments have sought to address transboundary environmental problems. Since the Treaty of Westphalia in 1648, the world has been guided by, and governments have sought to reinforce, international norms of state recognition, sovereignty and non-intervention. According to these prevailing and powerful norms, states are the ultimate and most legitimate expressions of human organisation, the venues for morality and the solutions to major challenges that extend beyond individual communities. These norms have so far largely guided discourse, thinking and responses to transboundary environmental problems: international environmental diplomacy, regimes and treaties have been based (almost by definition) on the responsibilities, obligations and capabilities of *states* to limit their pollution or use of resources, and to work together to cope with the effects of environmental harm and resource exploitation. The Westphalian international norms have been so powerful as to result in a doctrine of *international* environmental justice, manifested in the principle of common but differentiated responsibility among states. This doctrine has guided the creation of many recent international environmental agreements, but states have been noteworthy for the degree to which they have failed to implement it. This is a consequence of the doctrine itself. In the case of climate change, Westphalian norms have stifled diplomacy and prevented policy innovations, fundamentally ignoring the rights, responsibilities and duties of *individuals*.

Chapter 4 describes the international climate change regime and its provisions for international environmental justice. It outlines how the international response to climate change has failed adequately to address the problem. The doctrine of international environmental justice that has emanated from Westphalian norms, discourse and thinking has taken the world politics of climate change in a direction that has been characterised by diplomatic delay, minimal (or no) action – especially relative to the scale of the problem – and mutual blame between rich and poor states resulting in a 'you-go-first' mentality even as global greenhouse gas emissions skyrocket. The doctrine is one premised on national interests, which in practice routinely translates into national selfishness. The international doctrine has been written into international agreements such as the United Nations Framework Convention on Climate Change, the Kyoto Protocol and subsequent agreements and diplomatic negotiations on implementing the protocol and devising its successor. Although some major industrialised states in Europe have started to restrict and even reduce their emissions of greenhouse gases, these responses pale in comparison to the major cuts (exceeding 80 per cent, *at minimum*) demanded by scientists (Speth 2008: 29). Indeed, global emissions of greenhouse gases are *increasing*, and will do so for decades to come unless drastic action is taken very soon. This is in large part due to huge emissions increases being experienced in many major developing countries as their economies grow and as millions of their citizens adopt Western consumption patterns. At present, however, emissions from the expanding wealthy classes and new consumers of the world are excluded from the climate change regime because most of the states in which those people live are victims of pollution from traditional consumers in the world's wealthy countries. This exclusion obtains despite the growing impact of this new consumption and pollution on the Earth's atmosphere.

GLOBAL JUSTICE

One potentially potent remedy to the Westphalian norms that have plagued responses to climate change can be found in cosmopolitan ethics and global conceptions of justice that routinely and explicitly consider people as well as states. Chapter 5 defines cosmopolitanism and looks at what this perspective tells us about justice in a highly globalised world. A cosmopolitan approach places rights and obligations at the individual level and discounts the importance of national identities and state boundaries. Cosmopolitans recognise the obligations and duties of responsible and capable individuals regardless of their nationality. From a cosmopolitan perspective, what matters are (for example) affluent Americans and affluent Chinese *people*, rather than the 'United States' or 'China' qua *states*. People in one state do not matter more than people in others. Cosmopolitan justice makes demands on capable individuals for a range of reasons, such as the prescription to 'do no harm' (Shue 1995), the historical argument of 'you broke it, you fix it' (Singer 2003), the maxim to 'prevent extreme suffering' (Singer 2003), the belief in the 'ability to benefit others or prevent harm' (Jamieson 1997), the 'priority of vital interests' (Barry 1998) and the concept of not undermining others' capacity to be independent moral agents (O'Neill 1988). Generally speaking, *international* justice views national borders as being the basis for justice. In contrast, *global* justice, while accepting that national borders have great importance in the world, sees them as being the wrong basis for justice. This is especially so in the case of climate change.

Chapter 6 examines perhaps the most important development in the world today: the rise of hundreds of millions of new consumers in a number of developing countries. As recently as the late twentieth century it was possible to talk about climate change, in both practical and moral terms, by exclusively pointing to the responsibility of developed countries and their citizens as the causes of atmospheric pollution and as the bearers of duties to end that pollution, make amends for it and aid those who will suffer from it. The climate change regime, insofar as it recognises this responsibility, is premised on this notion. But the world has changed dramatically in recent years. The developing countries together now produce fully half of the world's greenhouse gases. China has overtaken the United States to become the largest national source of these pollutants. Given the developing countries' large populations, this change does not in itself alter the moral calculus very much because their national per-capita emissions remain low relative to those of the developed countries. What has changed, however, is the increasing number of new consumers in these countries, many of them very affluent indeed, who are living lifestyles analogous to, and often superior to (in terms of material consumption), most people in the developed countries. Now numbering in the hundreds of millions, these people are producing greenhouse gases through voluntary consumption at a pace and scale never experienced. While many societies in the West are finally starting to make changes that limit and reduce their greenhouse gas emissions, the new consumers are going in the opposite direction, with truly monumental adverse consequences for the atmospheric commons. At present, these new consumers face no legal obligations to mitigate the harm they do to the environment, and they have so far escaped moral scrutiny. If solutions to climate change are to be found, this will have to change, not least because 'old consumers' in developed societies will be watching these new consumers do the things that the old consumers are being told they must not do in order to help the world tackle climate change. As long as the new consumers hide behind their states' poverty, practical and politically viable solutions to climate change will be very difficult to realise.

Chapter 7 proposes an alternative to the status quo climate change regime, premised as it is on the rights and duties of states while ignoring the rights and duties of too many people. The chapter proposes that cosmopolitan aims should be incorporated as *objectives* of climate change diplomacy and policy. Because cosmopolitanism is concerned with individuals, it can help the world reverse the failed national and international policies that have contributed to the tragedy of the atmospheric commons. It can do this in part by addressing the lack of legal obligations for many millions of affluent people in developing countries to limit their greenhouse gas emissions in any way while still recognising that the world's affluent states, and indeed the affluent people within them, are even more responsible to do so. Cosmopolitan justice points us to a fundamental conclusion: that affluent people *everywhere* should limit, and more often than not cut, their atmospheric pollution, regardless of where they live. This points to a cosmopolitan corollary to the doctrine of interstate justice, one that explicitly acknowledges and acts upon the duties of all affluent people, regardless of nationality, to be good global citizens. The cosmopolitan corollary comprises a new form of international discourse, a new set of assumptions about what states and their citizens should be aiming for, and a new kind of institutionalism that folds world ethics and global justice into the practice of states. This corollary is more principled, more practical and indeed more politically viable than current doctrine and norms of international environmental justice applied to climate change.

The book concludes in Chapter 8 by briefly looking at the importance of global citizenship and personal responsibility for actualising global environmental justice. The cosmopolitan corollary to international justice offers an escape from the legal and mental straitjacket of

Westphalian norms. By associating the pollution of individuals and classes of people with ethical diplomatic arguments, international agreements and the domestic policies intended for implementation of those agreements, governments of both developed and developing states can escape the ongoing blame game in which poor states blame rich ones for the problem's creation, and rich states blame poor ones for the problem's future trajectory – with both refusing sufficiently to obligate even their affluent citizens to do all that is necessary and just. In the context of climate change, cosmopolitan justice has the potential to define a pathway whereby major developing-country governments can simultaneously assert and defend their well-justified arguments rejecting *national* climate change-related obligations while also acknowledging and regulating growing pollution among significant segments of their populations. This in turn can help to neutralise the reticence of most developed-country governments and their publics to live up to their states' obligations finally to undertake the major cuts in greenhouse gas emissions that will be required to limit future damage to the atmospheric commons upon which all states and all people depend. The cosmopolitan corollary can also help to free up new financial resources to aid those people most harmed by climate change. The conclusion we are left with is that global justice is almost certainly unavoidable if climate change is to be addressed effectively any time soon.

SOME CAVEATS ON THE CONTENTS

I will not be spending much time debating, as philosophers do, the merits of many different ethical perspectives. Following Simon Caney (2005b: 16–17) to some extent, while I do introduce a number of cosmopolitan thinkers and traditions of thought, I focus on arguments that shed light on or support a particular course of action. I do not undertake to present a review of literature on climate justice, although readers will get some of that from the book as a whole. As such, the book is meant to be both an introduction to climate change and related questions of justice, and an attempt to craft a proposal for more effective policy action. While philosophers may be reluctant to accept it, the concepts of right and wrong in the real world of international politics are seldom based on philosophical minutiae. Instead, they are based (by my reckoning, at least) on broad, relatively clear and straightforward concepts and arguments.

This is where international politics comes in. One aim here is to advocate a cosmopolitan theme for international politics. I will not be arguing strongly for global cosmopolitan democracy or world government – although they may be good ideas and ultimately what we need to combat climate change effectively (see Heater 1996). Instead, my aim is to show how cosmopolitan world ethics are *practical and politically viable* in the context of climate change, and how global justice is possibly the most *realistic* route to a new climate change regime that tackles this problem in the forceful way that is required. I follow Nigel Dower (2007: 5), drawing upon normative theory to give 'a defence of an ethic for individuals in which the global dimension of responsibility is significant'. This is, by definition, a cosmopolitan argument, albeit one applied to relations among people and among states.

The argument here is fundamentally a cosmopolitan one because it assumes and attempts to support the contention that taking all people everywhere to be of equal moral worth, and basing climate change policies within and among states on the premise that people are equal in this sense, are the best ways to break out of the too-little, too-late approaches to climate change that the world has mustered to date. I do not attempt to present in detail, nor do I defend strongly, any particular cosmopolitan response to climate change. I review the philosophical literature selectively to help establish the case for a new type of climate policy. This will come as a disappointment to some philosophers who would prefer a very carefully crafted argument free of contradictions. So would I, but in the real world of politics, and probably most of all in *international* politics, contradictions are the norm. If policy-makers, diplomats, parliamentarians, activists and not-so-activist people are persuaded by my formulation of an alternative approach to tackling global climate change, I will be successful. If philosophers also see merit in what is here, all the better. My point is that there is a rough convergence from the perspective of many ethical theories on what needs to be done differently, especially by individuals. This has important implications for international politics and for the policies of states.

My argument in favour of a more cosmopolitan approach to dealing with climate change is not meant to be an idealistic exercise or an act of imploring the world to come around to the view that all people will soon feel that they are global citizens or that states can be abandoned. Rather, this is an attempt to show that the most practical and politically viable approach to climate change – as well as the most principled one – is in fact one that actualises cosmopolitan ethics, and more often than not can be and should be premised upon those same ethics. Dower (1997: 561–2) points to three considerations in world ethics: theoretical, normative and what we might call practical (that is, the application of norms). While theory and normative prescriptions are invoked here, ultimately my subject is the latter: it is about *practical* ethics, not idealistic or utopian visions and hopes.

This book is also not a philosophical treatise on justice generally or environmental justice in particular. It is not an attempt to argue in favour of one definition of justice over another. And it is not an attempt to repeat or affirm particular arguments for or against climate change justice made by other writers, although there will be some of this. Instead, the book is an attempt to show that justice must be part of the response to climate change, as others and I have argued and as the climate change regime recognises, but that the way that justice has been included in the climate change regime has not helped to solve the problem. In particular, this book is about the failure of the discourse and the practice of *international* justice. It is about an alternative approach to justice, premised on cosmopolitan considerations, that can provide more principled, practical and politically agreeable solutions for mitigating and adapting to global climate change. World ethics and *global* justice are the most realistic foundations for climate change diplomacy and policies.

CONCLUSION

The bulk of literature on justice and climate change, and all related international legal instruments, speak of obligations of *states* to act (or not) to limit their emissions of greenhouse gases, or to act in ways to mitigate the effects of these emissions, and to assist poorer states to help them develop in less polluting ways. There is much less discussion – and what debate we have is largely among philosophers and activists, not diplomats – about the obligations of *individuals*. Increasingly, however, individuals matter: more and more of them who are not now subject to any climate-related obligations are able to afford lifestyles that lead to greenhouse gas emissions and climate change. This is especially true given the very rapid increase in the numbers of affluent people in the developing world, most prominently in China and India. As Bradley Parks and Timmons Roberts (2006: 345) remark, 'climate scientists can barely fathom a world in which the families of China and India will drive their own cars'. But that is exactly the world that is emerging. In China alone, hundreds of millions of people are quite rapidly adopting Western consumerist lifestyles. The climate change regime ignores this new reality. Thus a central theme of the chapters to follow is that something is lacking in today's climate change diplomacy and policy: a sensitivity to the moral and realistic imperatives of *global* justice that encompasses all people everywhere.

A crucial question is, who is obliged to act to address climate change? Henry Shue (1992: 385) argues that justice is fundamentally 'about not squeezing people for everything one can get out of them, especially when they are already much worse off than oneself. A commitment to justice includes a willingness to choose to accept less good terms than one could have achieved – to accept only agreements that are fair to others as well as to oneself.' It is well established that states have some obligations to implement climate justice. We can take that as a given, even as states usually fail to live up to those obligations. Many will argue that other actors, notably corporations and perhaps international organisations, also have obligations.[4] And there is another answer to the question: affluent individuals *everywhere*, even including those living in the poorest countries, are obligated to act. Here we find support from Caney (2005a: 770), who (rather unusually) argues that 'the burden of dealing with climate change should rest predominantly with the wealthy of the world, by which I mean affluent persons in the world (not affluent countries)'. It is not unusual to say that rich people in economically developed states have obligations, so more will be said about affluent individuals in the developing countries, which is something remarked on quite rarely. The present situation, whereby affluent individuals in poor countries are completely off the hook, directly (as are most people in affluent countries) and indirectly (unlike people in some European states, who must pay more for energy as part of those countries' early efforts to act on climate change), hardly fits Shue's conception – and many other conceptions – of justice.

Throughout the book two main critiques and two main proposals are put forward. The first critique is that of the state-centric myopia of the climate change regime. Because the regime is premised on the rights and (less so) the responsibilities of sovereign states, it has resulted in a tragedy of the atmospheric commons and a climate change regime that largely ignores the roles, rights and duties of persons. The second critique is of some new, very important and absolutely essential cosmopolitan interpretations of the climate change problem. These interpretations rightly invoke various forms of global justice as possible remedies for climate change (often moral remedies, but sometimes practical and institutional ones). However, they routinely do what cosmopolitans ought not to do: they discriminate by treating advantaged, capable and affluent people in different countries very differently. In particular, they tend to ignore the real-world causes of climate change – and the moral implications of that reality – by making demands on affluent people in developed countries while not making the same demands on affluent consumers in developing countries. This preoccupation with people in rich countries creates moral, practical and political problems for addressing climate change.

The two main proposals of the book emanate in large part from the reasons for these two critiques. The first proposal is for moral cosmopolitanism to help overcome the myopia of interstate doctrine. Moral cosmopolitanism puts people first, as should discourse and negotiations on climate change. Related to this, the second proposal is for what is called cosmopolitan diplomacy to overcome the you-go-first mentality of the international climate change negotiations by making global justice, and thus human rights, central to climate diplomacy and a key objective of climate policy. States will remain key actors, but they ought to take on a new role of being facilitators of global citizenship.

The existing system of international environmental governance, like international relations generally, is biased against – and indeed premised upon – *not* placing any obligations directly on people within state boundaries. To do otherwise would tend to violate state sovereignty, or at least the usual conception of it. But our preoccupation with narrow conceptions of international justice diverts attention and action exclusively to the national and interstate levels, when what is needed is simultaneous attention to localised and individual responsibility and action. The current solutions – international agreements – will not do enough to address, fundamentally, the current global trajectory of greenhouse gases. Without very substantial changes in behaviour at the personal level, climate change will probably be exacerbated. To be sure, these changes in behaviour are mostly required of people in developed states, which is consistent with what developing countries have been rightly demanding for a long time. But there will also have to be changes in behaviour among millions of affluent people living outside the developed countries. This is an important message that has been transmitted too rarely so far. The upshot is that by placing persons – and their rights, needs and duties – at the centre of climate diplomacy and discourse, more just, effective and politically viable (and palatable) policies are increasingly likely to be formulated and implemented.

NOTES

1. Throughout the book, the terms 'country', 'nation' and 'state' are normally used synonymously to refer to sovereign states (e.g. Canada, China), while 'the world' normally refers to all humanity.
2. Barry is playing on the title of Al Gore's book, *An Inconvenient Truth* (2006).

3. The affluent also have obligations to act to protect and to aid non-humans and the biosphere, but that is not something that is addressed here (see, e.g., Barkdull and Harris 1998; Midgley 2001).
4. For a taxonomy of who or what should bear these burdens, see Caney (2005a: 754–5).

PART I
THE CHALLENGE

CHAPTER 1

GLOBAL CLIMATE CHANGE

Industrialisation, traditional economic development and modern life-styles are contributing to pollution that is warming the globe. This global warming is in turn causing changes to the Earth's climate system that are increasingly impacting on individuals and societies, especially in the poorest parts of the world. Tragically and unjustly, climate change will cause the most suffering among those least responsible for it. While most wealthy countries and people will be able to cope with climate change, at least for now, millions – and probably billions – of the world's poor will not be so lucky. They will not be able to avoid suffering from droughts, floods, severe cyclones, water shortages, crop failures and spreading pestilence. Historically, people in the world's affluent countries have been the main polluters of the atmosphere, often as a consequence of voluntary, and frequently frivolous, material consumption. Now the rapidly expanding middle and wealthy classes of the developing world are starting to do the same, with predictable harmful consequences for the environment. Thus climate change not only presents major practical challenges for individuals, societies and governments, but also raises the most profound questions of international and global justice yet encountered in human history.

This chapter establishes the scientific basis for discussions of these justice-related questions in subsequent chapters. I briefly introduce the causes of climate change before summarising some of the significant challenges that it presents for the world, now and in the future.

CAUSES OF CLIMATE CHANGE

In general terms, 'climate change' refers to changes in the Earth's climate resulting from global warming.[1] Human-induced global warming is caused by the build-up of greenhouse gases in the atmosphere and ecosystem. Greenhouse gas pollution results from human activities,

17

which in general terms fall into two categories: those that result in greenhouse gas emissions and those that destroy carbon 'sinks'. The most significant greenhouse gas in aggregate is carbon dioxide, which is released by the burning of fossil fuels (for example, coal, oil, natural gas) for transport, industry and domestic energy use, and through land-use changes, such as when forests are cut or burned. Carbon sinks are features of the ecosystem, such as trees, which absorb carbon dioxide. Other sinks include those for methane, a very powerful greenhouse gas – as, for example, the retention of methane in the vast tundra of Siberia. Greenhouse gas emissions and the destruction of sinks can occur together, such as during the cutting of tropical forests (destruction of carbon sinks) and then burning the felled trees and brush (emitting carbon dioxide) for cattle grazing (resulting in emissions of methane from grazing ruminants).

Over the last few decades, scientists have radically improved their understanding of the causes and consequence of global warming. The most authoritative scientific reports have come from the Intergovernmental Panel on Climate Change.[2] The panel, created by governments in 1988, has concluded with '*very high* confidence that the global average net effect of human activities since 1750 has been one of warming' (IPCC 2007a: 37). According to the panel's 2007 fourth assessment report, since the start of industrialisation in the eighteenth century, atmospheric concentrations of carbon dioxide, methane, nitrous oxide and other greenhouse gases have increased, primarily as a consequence of fossil-fuel burning, but also because of agriculture and other land-use changes. Since 1970 human-caused greenhouse gas emissions have increased globally by 70 per cent, with carbon dioxide increasing by 80 per cent since 1995 (IPCC 2007a). The panel reports that 'atmospheric concentrations of CO_2 [carbon dioxide] and CH_4 [methane] in 2005 exceed by far the natural range over the last 650,000 years' (IPCC 2007a: 37). The concentration of carbon dioxide in the atmosphere in 2005 was 379 parts per million (ppm) compared to 280 ppm prior to the Industrial Revolution, with the annual increase being nearly 2 ppm. Importantly, although plants and the oceans absorb carbon dioxide, global warming inhibits their ability to do so, thereby creating a positive feedback loop contributing to more warming and greater climate change.

Perhaps seeking to counter the political influence of 'climate sceptics' – who question the reality of global warming and attribute it to all manner of causes, such as sun spots – the intergovernmental panel has declared that 'warming of the climate system is unequivocal, as is now evident from observations of increases in global average air and ocean temperatures, widespread melting of snow and ice and rising global average sea level' (IPCC 2007a: 30). What is more, the panel found that 'discernible human influences extend beyond average temperature to other aspects of climate, including temperature extremes and wind patterns' (IPCC 2007a: 40). That is, most of the causes of climate change, and thus most of the consequences of it, are undoubtedly attributable to human activities.

The Intergovernmental Panel on Climate Change has considered the influence of planned and likely national sustainable development policies and efforts to mitigate climate change. Its findings are not optimistic. The panel projected that that there will be an increase of 0.2 degrees Celsius per decade under most emissions scenarios, although future temperature increases will depend on the world's responses. Global average temperature was predicted to rise by 1.4 to 5.8 degrees Celsius this century, with the highest increase likely if there are no additional mitigation policies. Even with anticipated policies for mitigating climate change (and related sustainable development policies), global greenhouse gas emissions will grow for at least several decades. Consequently, 'we have passed the point when the spectre of unacceptable climate impacts could still have been avoided . . . For the next decades, we are locked-in to an unavoidable rise in global mean temperature by virtue of our past emissions . . . This is unlikely to pass without creating serious climate hazards' (Muller 2002: 3).

To be sure, many human activities that contribute to climate change are essential. In most places, burning some type of fossil fuel is essential for survival or at least comfort, and in much of the world people have little choice but to use vegetation to provide fuel for cooking and heating. This suggests that population is a key driver of climate change. But, to be fair, we should focus not on population per se but on industrialisation and modern consumption choices. As the developed countries have demonstrated through most of their citizens' very high levels of pollution, it is not how many people are living that matters for climate change as much as how each of them lives, and particularly how much each of them consumes. Most people in developed societies and the expanding affluent minorities in developing countries live highly consumptive lifestyles that result in very high per-capita emissions of greenhouse gases. Thus, *modern lifestyle* is the major cause of climate change.

An example of this dynamic can be found in what people choose to eat. One major growing contributor to greenhouse gas emissions is the rapidly increasing consumption of meat, which is expected to double by 2050 (FAO 2007). Most of the increase will result from growing consumption in developing countries. Meat production requires prodigious amounts of land (to grow feed or for grazing), water (to produce animal feed and flush wastes) and energy (for production and transport), and many livestock animals, notably ruminants (for example, cattle and sheep), are sources of methane, a greenhouse gas twenty-three times more powerful (molecule for molecule) than carbon dioxide. Incredibly, meat production produces more greenhouse gases than does the entire transport sector, accounting for about 18 per cent of all human-caused greenhouse gas emissions worldwide (FAO 2007).

CONSEQUENCES OF CLIMATE CHANGE

Human-induced global warming was, until quite recently, perceived to be a future problem. But it has become clear that *ongoing* climatic changes are consequences of global warming (*New Scientist* 2006). The world has warmed substantially in recent decades, with most of the years since the mid-1990s setting records for being the hottest since record keeping began in the mid-nineteenth century. Natural ecosystems are *already* being affected by regional temperature increases and other climatic changes. The impacts of climate change on these natural ecosystems and indeed on human society and economies are potentially very severe, particularly in parts of the world where geographic vulnerability and poverty make adaptation difficult or impossible. Among many ongoing adverse impacts of climate change, the proportion of the Earth affected by drought has increased, as has the frequency of extreme weather events, heavy precipitation, intense tropical cyclones, extremely high sea levels and heat waves (in most regions). In recent decades the number of people around the world affected by weather-related disasters has quadrupled to reach hundreds of millions of people, with predictions of those affected by 2030 rising to numbers approaching one billion, far higher than any recorded disaster in the latter part of the twentieth century (Muller 2002: 4). Meanwhile, the frequency of cool days and nights has declined. These changes are having noticeable effects on both physical and biological systems, as demonstrated by melting glaciers and sea ice; warming of lakes and rivers; the early advent of spring and associated changes to plants and wildlife, such as the earlier greening of vegetation and the corresponding impacts on bird migration and egg laying; and major alterations in marine ecosystems, including changes in salinity and currents, ranges of marine life, timing and locations of fish migrations, probable adverse impacts on reefs, and losses of coastal wetlands and mangroves (both crucial for healthy fisheries). The Intergovernmental Panel on Climate Change also reports adverse changes to agriculture and harm to forests from more fires and pests. Human health has also been affected by heat stresses and expanding ranges of disease vectors (for example, mosquitoes), among other effects.

Specific *future* impacts this century on environmental systems and human sectors predicted by the intergovernmental panel include the following: freshwater resources will change, with water runoff by mid-century likely to increase substantially in some high latitude and tropical areas while decreasing in dry areas already suffering from water shortages. Areas affected by drought will expand, whereas flooding from heavy precipitation will increase. Water stored in glaciers and snow cover will decline, reducing water supplies to much of the world's population. Ecosystem resilience will probably be surpassed owing to flooding, drought, fires, insects and ocean acidification, with 20–30 per cent of plant and animal species *already* considered at risk of extinction this century. If temperature increases exceed 1.5–2.5 degrees Celsius, the negative consequences for biodiversity, water and food supplies, as well as other ecosystem services, will be major. Crop productivity may increase at first in some higher latitude areas but then decline as regional temperatures increase; declines are expected in tropical regions and seasonally dry areas even with small temperature increases. World food production is predicted to decline if global temperature increases exceed 3 degrees Celsius. Increasingly frequent floods and droughts will adversely affect local crop production, especially in low-latitude regions where subsistence agriculture is the norm. Fish species are likely to move to new areas, with fisheries and aquaculture predicted to be adversely affected. Coastal ecosystems and wetlands (for example, salt marshes and mangrove forests) will be damaged by erosion and sea-level rise. Coastal flooding will increase, resulting in severe effects for people in densely populated low-lying areas, notably small-island countries and mega-delta regions of Africa and Asia. Coral reefs and species dependent on them will be harmed by increasing ocean temperatures and by ocean acidification. Overall net effects of climate change for industry, human settlements and societies are projected to be negative, with the adverse impacts increasing as temperatures increase. This is especially true of industries, settlements and societies in coastal areas and in river flood plains and in areas where people are heavily reliant on resources impacted by climatic changes – meaning the poorest places with the poorest people.

The effects of climate change, especially in poor places, will be increasingly harmful to the health of millions of people. Projected health effects include increased malnutrition (and related disorders, particularly for children); more deaths, diseases and injuries caused by heat waves, fires, droughts, storms and floods; increased diarrhoea and harm to health from climate-related air pollution (ground-level ozone); and diseases resulting from the spread of infectious pests to wider areas. Even in affluent parts of the world that have greater aggregate capacity to adapt, some groups of people, notably the poor and the elderly, will suffer the risks of climate change. The upshot is that, around the world, far more people are projected to be harmed than helped by climate change, even if temperature increases are somehow mitigated (Working Group II 2001: para. 2.8).

REGIONAL EFFECTS OF CLIMATE CHANGE

The intergovernmental panel has predicted the impacts of climate change for this century in specific regions of the world. Africa is among the continents most vulnerable to climate change, both because of the impacts themselves, which are multiple, but also because in most of Africa the ability to adapt to those impacts is generally very low. Climate change-related water stress in Africa is likely to affect up to 250 million people, undermining livelihoods and making existing water problems worse. Access to food is forecast to be 'severely compromised' (IPCC 2007b: 13), agricultural yields will decline in some areas by up to 50 per cent, areas suitable for agriculture will be reduced, and the length of growing seasons will shorten, exacerbating African malnutrition and lowering already low food security. Fisheries in Africa's lakes are likely to decline owing to increased water temperatures, and sea-level rise along Africa's coastlines will adversely affect mangroves, coral reefs and nearby fisheries. Adaptation to sea-level rise could cost at least 5–10 per cent of gross domestic product.

In Asia, economic development will suffer as climate change exacerbates existing pressures on the environment and natural resources resulting from industrialisation, economic growth and urbanisation. Flooding will at first increase in areas dependent on glacial melt waters, but as glaciers recede river flows will decrease. Fresh water in most of Asia will become less available, possibly affecting more than one billion people by mid-century. Flooding from the sea, and in some mega-delta regions flooding from rivers, will affect many heavily populated areas. While crop yields may increase in East and South East Asia, in other parts of Asia they could decrease by 30 per cent or more, with the overall risk of hunger in the region expected to remain very high. Diseases, especially those related to floods and droughts (for example, diarrhoea) will increase morbidity and mortality, and rising coastal water temperatures could make other diseases more potent (for example, increased cholera toxicity).

   In Australia and New Zealand governments have substantial capacity
to adapt to climate change, but even there many of the impacts will be
too extreme to cope with, and natural systems and species will often be
unable to adapt, leading to significant biodiversity losses. Vulnerability
was demonstrated in Australia in early 2009 when ferocious bush fires
killed hundreds of people. Water security will be a major concern in
large parts of these countries owing to less precipitation and more
evaporation. Sea-level rise, as well as increasingly severe and frequent
tropical storms, will increase risks to developed and populated coastal
areas. By 2030 agricultural production will decline over large areas
(although there will be temporary increases in some parts of New
Zealand as growing seasons lengthen).

   Like Australia and New Zealand, the countries of Europe are also
well equipped to adapt to the impacts of climate change, but the costs of
doing so will be high and 'nearly all European regions are anticipated to
be negatively affected by some future impacts' (IPCC 2007b: 14). Future
climate-related changes across Europe are likely to mirror many ob-
served changes in recent years: glaciers retreating, species moving
northward and up mountain slopes, and unprecedented heat waves.
Negative impacts will include more flash flooding from rivers, coastal
flooding and erosion (from storms combined with sea-level rise), reduced winter snow cover (harming tourism) and 'extensive loss' of
species (up to 60 per cent in some areas under probable greenhouse
gas emissions projections) (IPCC 2007b: 14). Southern Europe will
experience worsening high temperatures, drought and falling agricultural productivity. Central and Eastern Europe will experience increased
water stress and falling forest production owing to less rainfall, and
increased health risks from heat waves. Northern Europe will see more
mixed effects, including temporary increases in crop yields from longer
growing seasons, although negative impacts will eventually outweigh
any benefits as the region experiences more winter floods, landslides and
degraded ecosystems.

   Latin America includes some countries relatively well equipped to
adapt to climate change but many others that are not, the latter often
because of low levels of economic development. One major impact of
climate change in tropical parts of this region will be significant
extinction of species, with tropical rainforests gradually being replaced by savannah in eastern Amazonia. In areas that are already
dry, conditions will grow worse, with salinisation and desertification
of agricultural land. Crop productivity will fall in many areas.
Availability of water will fall, owing to reduced precipitation and
disappearing glaciers, leading to scarcities for human consumption,

hydropower and croplands. Low-lying coastal areas will suffer flooding from sea-level rise, and rising sea temperatures will harm reefs and fisheries.

The countries of North America are more able to adapt to climate change than are most parts of the world, but major vulnerabilities exist, as Hurricane Katrina's impacts on New Orleans vividly demonstrated in 2005. Water scarcities will increase in western mountain regions as snowpack decreases, summertime river flows fall and winter flooding increases. Forests will suffer from large increases in fires, pest infestations and disease. While rain-fed agriculture will increase in some areas for a period, in the many areas where irrigation water is already heavily utilised crop yields will suffer. Cities now experiencing heat waves will see them increase in frequency, duration and intensity, with adverse impacts on human health, especially for the elderly. Coastal areas, including local communities and coastline habitats, will experience the combined harmful effects of climate change impacts, development and pollution. Because the value of infrastructure and built environments in these areas is increasing, there will be growing economic losses from storms exacerbated by sea-level rise.

Polar regions will experience loss of glaciers and ice sheets, as well as detrimental changes in natural ecosystems (for example, migratory birds and mammals). The Arctic will see the extent and thickness of sea ice and permafrost fall, and the amount of coastal erosion will increase. Human communities will be affected as infrastructure is damaged and traditional ways of life become impossible because of changing snow and ice conditions.

Small islands are especially vulnerable to climate change, particularly in the case of low-lying islands that will be inundated by sea-level rise and storm surges that will cause extensive damage to coastal infrastructure and settlements. Coastal erosion and coral bleaching will increase, adversely affecting fisheries and tourism. Freshwater resources will be reduced with climate change to the point where they will not meet demand in periods of low rainfall. Invasion of non-native species will increase on many small islands as temperatures increase.

Overall, the effects of climate change later this century, as predicted by the intergovernmental panel, are usually negative and frequently severe. With continued warming, expected manifestations of climate change will be 'larger' (that is, usually more adverse) than those seen in the last century (IPCC 2007a: 45). As a consequence, 'the resilience of many ecosystems is likely to be exceeded this century by an unprecedented combination of climate change, associated disturbances (for example, flooding, drought, wildfire, insects, ocean acidification) and

other global change drivers (e.g. land-use change, pollution, fragmenta-
tion of natural systems, overexploitation of resources)' (IPCC 2007a:
48). 'Positive feedbacks' will increase as carbon uptake by plants reaches
saturation. The likelihood of abrupt or irreversible environmental
changes will increase, with some of them considered inevitable. These
could include rapid sea-level rise, significant extinctions (40–70 per cent
of species if temperature increases exceed 3.5 degrees Celsius), large-
scale, persistent changes to marine systems and fisheries, and yet more
positive (that is, harmful) feedback loops as oceans lose their ability to
absorb carbon dioxide. The upshot is that the effects of climate change
'will vary regionally but, aggregated and discounted to the present, they
are very likely to impose net annual costs which will increase over time
as global temperatures increase' (IPCC 2007b: 17). If temperatures
increase beyond modest predictions – if they exceed 2–3 degrees
Celsius (as they almost certainly will) – even the few positive benefits
for some regions will be very brief. And if we look beyond this century
the impacts of climate change are expected to be much more severe
than those described here. They could be, and probably will be,
monumental.

POVERTY AND CLIMATE CHANGE

An important theme is apparent from accounts of the impacts of climate
change: the relationship between climate change-related suffering and
poverty is decidedly direct; as climate change increases, so too does the
poverty of poor countries and poor people, and as climate change
increases so too does suffering of the poorest countries and people. For
example, climate change will probably cause extreme suffering in southern Asia and southern Africa owing to falls in essential crop yields (Tin
2008: 4). According to Dupont (2008: 33), by 2025 half a billion people
could be suffering from serious water shortages caused by climate
change. Those few areas projected to experience some temporary
benefits from climate change, such as longer growing seasons, are most
often in already well-developed regions and countries that are best able
to adapt (for example, Europe, North America and New Zealand). Poor
people and communities are the most vulnerable to 'natural' hazards
because they rely heavily upon environmental resources, often lack
secure access to those resources, do not have the assets necessary to
rebound following environmental upsets, and routinely do not have
access to institutional resources that help to limit the impacts of
disasters and to cope with harmful natural events (Smith 2006: 8–9).
Thus those people who will suffer the most from climate change are
those who already suffer from poverty and destitution, and many of

those who are barely out of poverty could be pushed back into it as a consequence of climate change.

Climate change increases poverty because it threatens the full range of capital necessary for livelihood: physical capital is undermined by damage to infrastructure and homes; human capital is harmed by infectious disease and injuries from extreme weather; social capital is weakened when people are forced to move away from affected areas; natural capital is degraded with loss of biodiversity, agricultural productivity and water scarcity; and financial capital is lost when the effects of climate change, such as floods and droughts, reduce income and undermine economic development (Smith 2006: 27). By their very nature, of course, those people who will suffer the most from climate change are also those who have contributed, and are now contributing, the least to it.

We will not have to wait long for the more severe consequences of climate change to be felt. Recently it has become very clear that the intergovernmental panel's predictions for this century have been overly optimistic. Things are very likely to be much worse than just described – within the lifetimes of most readers of this book.

CONCLUSION

Largely because of improving expert knowledge of climate change and the increasingly dire predictions by scientists of its effects, governments have slowly come to realise the dangers it poses. They have recognised that addressing climate change requires collective action. While states are not the proximate causes of climate change (after all, a state exists only as a set of institutions based on certain ideas), and while we know that the actions of people and machines are what actually cause climate change, the problem has been viewed as something for states (that is, governments) to work together to investigate and to solve. (Even the overarching scientific organisation that has become the premier authority on climate change is the *Intergovernmental* Panel on Climate Change.) Solutions to climate change within borders have primarily been sought in government policies (for example, regulations and taxes). Beyond borders the focus has been on interstate diplomacy, conventions and protocols, with the most obvious results being the 1992 Framework Convention on Climate Change (henceforth referred to as the climate change convention) and the 1997 Kyoto Protocol to that convention, as well as a number of other associated agreements and ongoing negotiations that we can collectively refer to as the climate change regime (see Chapter 4).

The primary objective of the climate change convention is to stabilise the concentration of greenhouse gas concentrations in the atmosphere at a level where they will not result in dangerous changes to the Earth's climate system (UNFCCC 1992: art. 2). To achieve this objective, governments agreed that climate change is a *common but differentiated responsibility*: all states are responsible for addressing climate change, but the affluent ones, which are the largest historical polluters of the atmosphere, agreed that they were obligated to act first to reduce their emissions of greenhouse gases before the developing countries were required to limit theirs. Some governments have started to act on their obligations, as reflected in efforts by some European states to limit their greenhouse gas emissions (Harris 2006). However, these efforts have been tiny compared to what is required. By almost every measure, climate change is already shaping up to be extremely dangerous, contributing to environmental damage and human suffering now and much more so in the future, especially in the poorest parts of the world. Indeed, as we will see in Chapter 4, global warming and climate change are more severe than even the worst-case predictions of the Intergovernmental Panel on Climate Change. In short, the international climate change regime is failing (Harris 2007a).

Insofar as justice has been part of the climate change regime, it has been conceptualised as international – that is, *interstate* – justice, and then mostly in terms of positive international law whereby states negotiate and agree to the terms and limits of justice, rather than accepting and fully codifying justice based primarily on moral norms. By diverting all responsibility to states, focusing on international justice has not discouraged consumption and pollution by affluent *people*, including both the hundreds of millions of polluters in the wealthy countries, where some controls on greenhouse gas pollution are being put in place in accordance with the climate change agreements, and the many millions of recent new consumers in developing countries, whose state governments have no obligation to limit nationwide greenhouse gas pollution. This raises the general question of whether justice does or should extend beyond borders at all, and if so to whom and how. It is to this question that the next chapter turns, as it does to the question of how we can define justice, particularly in the context of climate change.

NOTES

1. 'Climate change' in the lexicon of the Intergovernmental Panel on Climate Change refers to changes from both natural processes and human activities,

    whereas the Framework Convention on Climate Change focuses on the latter (see below).

2. This section and the next summarise some findings of the multi-volume fourth assessment report of the Intergovernmental Panel on Climate Change as recounted in IPCC (2007a, 2007b, 2007c, 2007d).

CHAPTER 2

JUSTICE IN A CHANGING WORLD

Before we focus on questions of climate ethics and especially climate justice, it is worth pausing to ask to what extent ethics and justice are germane to world affairs generally. The question leads to different answers depending on to whom it is asked (cf. Dower 1997: 563–6). Some people, sometimes labelled 'sceptical realists', believe that ethical norms and justice have little or no place in world affairs. These so-called realists believe that states are the key actors in world affairs, and that their relations are dictated by power and calculations of national interests. Insofar as ethics play any role, it is in the state's obligation to defend its own ethical space from interference. Other people, who are sometimes called 'internationalists', believe that ethical norms routinely obtain in world affairs, but only among states. From this 'morality-of-states' perspective, ethical norms exist within a 'society' of states as a means to maintain order and thereby protect national interests. According to this view, human rights and values can best be promoted by focusing on creating order among states. In practice, the extent to which justice obtains across borders in relations among states and other actors depends on the issue at hand, but generally speaking justice beyond borders has been applied to more and more issues in recent decades. When it has not been applied, ethical prescriptions frequently demand that it should be.

COMMUNITARIANISM VERSUS COSMOPOLITANISM

Discussions of justice in world affairs frequently take either a communitarian or a cosmopolitan perspective. Each perspective emphasises different rights and obligations. As the label suggests, communitarians would probably say that obligation is close to oneself – to one's family, neighbours and nation. Communitarians emphasise that individuals are

29

constituted, at least in large part, by the communities in which they live, and that assessments of what is just will derive from people's lives within their own communities (Thompson 1992: 21–2). At the root of communitarianism is a belief that 'value stems from the community, that the individual finds meaning in life by virtue of his or her membership in a political community' (Brown 1992: 55). According to this viewpoint, to the extent that people have a moral obligation to one another, they need act to help people only in their own community or at most their own country (if community and country are not the same). To be sure, people in the rich countries may recognise that they can promote their own interests by providing aid to poor countries, or they may choose to be charitable towards them. But a typical communitarian would usually argue that the rich states and their citizens have no *duty* of justice towards the governments of poor countries, nor do they have any *duty* towards individuals in those countries. Insofar as there is such a thing as international society, a communitarian might argue, it is not analogous to domestic society. Consequently, it would follow that we should not apply principles of justice very widely in international relations, least of all in wider global affairs.

These communitarian notions are crucial to our understanding of climate change and the failure of states to address it adequately. Having said that, subsequent chapters will not devote considerable time to communitarian thinking per se because it is a view that has essentially won the battle of ideas, and indeed has prevailed for centuries. That is, the interstate system under which we live today is one based on communitarian principles, often in the extreme. This Westphalian world view (see below and Chapter 3) is one premised on a particular kind of communitarianism, which asserts that people's identities and their moral values arise not from some common humanity or from universal values, but rather from shared traditions within established communities (cf. Dower 1997: 561). According to this view, morality varies from place to place; values that are important and widely endorsed and exercised in one place need not be so in other places, nor should they be judged from the perspective of universal norms or values. From this perspective, universal values are not inherently significant, and no global obligations arise – apart, perhaps, from respecting state sovereignty – from them. Given that this way of thinking – this international or interstate doctrine – has not done much to help the world to deal effectively with climate change, and arguably is a reason we have it, I will focus on an alternative, cosmopolitan approach to the problem, albeit one intended to modify communitarian, state-centric approaches rather than to replace them outright. I will call this the cosmopolitan corollary to interstate (communitarian) doctrine (see Chapter 7).

In challenge to communitarian notions, cosmopolitans believe not only that ethical norms apply to world affairs, but that they specifically apply to all people, regardless of the communities and states they might call home. They see obligation extending far away, even to people whom we might never see or hear about, let alone meet in person. 'Realists' and internationalists claim to describe the world and to base their arguments about the role of ethics – or lack of a role, according to 'sceptical realists' – on a world that they believe exists. In contrast, while many cosmopolitans observe features of a global community in which people are 'world citizens', their project, and that of many cosmopolitans who do not yet observe a global community as being in place, is not simply to describe the way the world is but to find ways of justifying, advocating and actualising a global community premised on and promoting human rights and mutual obligations. In this sense cosmopolitanism is often seen to be a normative world view. From this perspective, 'we are global citizens and have duties . . . whether most of us accept this now or not. National boundaries do not . . . have any ultimate significance' (Dower 1997: 565), meaning that one could potentially have multiple loyalties, including to one's own state but also to other human beings living far away, and even to other species. While cosmopolitans differ among themselves about what constitutes justice – as do communitarians – they usually share the conviction that any social order must derive its justification from how it affects the welfare of individual persons, assessed apart from the social relationships of those individuals, and 'urge that we live in a world governed by overarching principles of rights and justice' (Vertovec and Cohen 2002: 10). Thus, cosmopolitanism is ' "universalist" and "totalising" ' (Thompson 1992: 21; see also below Chapter 5).

A cosmopolitan might argue that every person whose basic needs have been met has some obligation to help those whose basic needs have not been met. Peter Singer (1996: 28) has suggested that such obligations extend far and wide: 'If we accept any principle of impartiality, universalisability, equality, or whatever, we cannot discriminate against someone merely because he is far away from us (or we are far away).' Charles Beitz (1979a) has said that cosmopolitan standards of international distributive justice, specifically egalitarianism, exist because all persons should be treated fairly regardless of their nationality. As Onora O'Neill (1986: 282) has argued, 'if complex, reasoned communication and association breach boundaries, why should not principles of justice do so too?' At the very least, a cosmopolitan might say, we have an obligation not to deprive people in other countries of their basic subsistence needs, because to do so would violate all of their other rights for which survival is a prerequisite (cf. Shue 1996a). Even if governments will not help each other, people ought to – or at least they have a moral obligation to do so and ought to try to do so. To use Stanley Hoffman's (1981: 153) words (which are not his own sentiments), 'to put it bluntly, our obligation of justice toward the Bantus is exactly the same as our obligation of justice toward immediate neighbors'. Even if we are psychologically incapable of fully comprehending the plight of people far away, we still have a moral obligation to assist in their welfare.

DEFINING JUSTICE

Justice obtains in world affairs to varying degrees, ranging from mostly within sovereign states (near the communitarian end of a spectrum), to widely among all states, and on to among all people (at the cosmopolitan end of a spectrum). But what do we mean by 'justice'? It may be neither possible nor desirable to have only one definition of justice; governments and individuals may never agree on one and to do so might stifle consideration of inequities that would not fit into that definition. Indeed, to expect to find a single definition of justice may be a 'hopeless and pompous task' (Tornblom 1992: 177). When surveying the literature on social justice, as Bernard Cullen (1992: 60) notes, 'the dominant impression is of something approaching philosophical pandemonium. When researchers in other disciplines . . . concerned with issues of justice and injustice approach the philosophers in search of a generally accepted definition and analysis of the object of their study, they are faced with a cacophony of discordant philosophical voices. Probably the most apt term to characterize the dozens of theories . . . is "incommensurability".' What constitutes justice even *within* national societies is subject to profound disagreement, even in the most economically developed societies. Such is also the case with applications of justice beyond state borders.

Despite the difficulties, it is possible to point to notions that commonly fall under the rubric of justice. Generally speaking, justice refers to the idea that individuals ought to receive the treatment that is proper and fitting for them – in short, to each his or her due. Justice is about identifying to whom rights are owed and to whom associated duties should be assigned and how much of the burdens of protecting those rights each actor with duties should bear (Jones 1999: 5–6). At the core of justice is the idea that 'no one should be preferred if others are thereby put at a disadvantage, and that no one should be harmed for someone else's advantage' (Sachs and Santarius 2007: 125). Three principles are often put forward to achieve just outcomes: 'everyone is to be taken into account in accordance with their rights, their needs or their performance. The conflict among these principles . . . is in large part the substance of struggles over justice' (Sachs and Santarius 2007: 129).

One important consideration for justice is the manner in which benefits and burdens are distributed among human beings. This book, like the international climate change regime it describes, and indeed the alternative regime it advocates, is most concerned with social (as opposed to legal) justice or, more specifically, distributive justice, meaning the 'fairness' or 'rightness' of distributing benefits and burdens (although legal justice will be important later in the book when we look at how it manifests itself as part of interstate doctrine in international environmental agreements). This still leaves the question of precisely what is fair and right. Specific attributes of justice are, for better or worse, often the result of political bargaining among interested actors. Often the concern is with distributing economic benefits, but frequently the goal is to distribute fairly other things that people care about, such as political power. Justice is not concerned merely with who has how much of what. For one person to have more than another is not intrinsically unjust. For us to apply conceptions of justice to any given situation, we usually look for some relation between actors that somehow affects the distribution (Arthur and Shaw 1978: 2–8). Thus, social justice is normally concerned with distributing benefits and burdens that are a result of social relationships and institutions (D. Miller 1976: 22).

Beyond the notion of distribution within some sort of social arrangement, what else might we mean by justice? David Schlosberg (2007: 11–41) indentifies a number of conceptions of justice proposed by political theorists in recent decades, including those focusing on distribution (or equity), recognition, capabilities and procedure. We might start by saying that each person ought to receive his or her due, based on rights, equality, fairness, merit (desert), need or some other criteria, or that each person ought to receive a share of some good (or not suffer some 'bad') depending on the extent to which that distribution will contribute to some desired consequence (for example, the utilitarian would seek to promote the greatest overall 'happiness'). Discussions of justice frequently involve two general categories of issues: procedural (dealing with how policies are decided) and consequentialist (the outcomes of decisions and actions). Procedural justice requires that rights of actors be respected in decision making, and that those affected by decisions be allowed to participate in the formulation of those decisions. Thus we might say that those affected by pollution ought to have a say in how it will be prevented or mitigated, as well as being involved in deciding who will benefit from efforts to make right past pollution. Consequentialist justice demands that there be a fair (however defined) actual distribution of burdens and allocation of benefits. Past polluters might be required to pay those who have suffered from their pollution. From the perspective of outcomes or consequences, we might ask: 'Who, and how many, should have how much? What would a fair distribution look like? Would it encompass striking inequalities?' (Almond 1995: 12).

Global environmental change can have a profound effect on interpretations of justice. David Miller (1976) points out that for a state of affairs to be unjust it must result from the actions of persons, or at least be capable of being changed by human actions. He goes on to illustrate this point by example: 'although we generally regard rain as burdensome and sunshine as beneficial, a state of affairs in which half of England is drenched by rain while the other half is bathed in sunshine cannot be discussed (except metaphorically) in terms of justice – unless we happen to believe that Divine intervention has caused this state of affairs, or that meteorologists could alter it' (D. Miller 1976: 18). It is ironic indeed that such a discussion would be hardly metaphorical today. To put it bluntly, today we *are* altering the weather and the Earth's climate. Industry and over-consumption on one side of the Earth cause foul weather on the other side. Foul weather in England may mean heavy rain. In sub-Saharan Africa it could mean lack of rain. In both cases it may be caused by what people do halfway around the world. Miller's example of where justice does not apply is precisely where it does in current and especially future contexts of global environmental change. Indeed, according to Steve Vanderheiden (2008: p. xiv), the 'global nature' of climate change 'defies conventional assumptions about state sovereignty and the geographically bounded nature of principles of justice. By the nature of individual GHG [greenhouse gas] emissions, the conventional assumptions regarding moral and legal responsibility are complicated by the complex causal change and aggregative nature of climate-related harm . . .'.

Thus climate change and international collaboration to deal with it and other environmental changes pose profound burdens and potential benefits for almost all countries and people, presenting us with important practical and ethical questions of justice. *International* justice, the subject of Part II, 'takes political communities organized into states to be the main agents of justice (i.e. who is asked to be just and who receives just treatment)', whereas *global* justice, the subject of Part III, 'takes persons, regardless of their political membership, as the primary focus of justice' (Forst 2001: 160). The former aims to 'regulate the relations between states in a fair way', whereas the latter sees principles of justice as existing 'to regulate relations between all human beings in the world and to ensure their individual well-being' (Forst 2001: 160). In the context of climate change, like almost all other areas of international politics, the international or 'statist' perspective has prevailed, in large part because the climate change regime has been negotiated and created by states on behalf of states and their perceived individual and collective interests. Consequently, the discussion in the coming chapters takes *international* justice as the default or reference position (insofar as justice obtains beyond borders) from which we can assess extant climate policies and look for more effective future policies.

ENVIRONMENTAL INJUSTICE AND CLIMATE CHANGE

As Nigel Dower (2000: 267) notes, 'ecology is perhaps the most obvious area for global ethics', given the way that environmental problems have taken on a global dimension. Global warming and resulting climate change are the most profound manifestations of this transformation in ecology. Climate change is increasingly an inescapable phenomenon for everyone, and its adverse impacts are increasingly unavoidable for the world's least well-off people. These increasingly severe impacts arising as a consequence of climate change mean that it is a profound matter of (in)justice, a situation that generates associated duties for those causing the problem. Environmental injustice is in part about 'the inequities – whether distributive or procedural – and disproportionate burdens borne by poor communities and countries that arise from the inclusion of what arguably are morally irrelevant features such as race, ethnicity, economic status and political power, in environmental decision making' (Hayden 2005: 136). When defined by affected communities, environmental justice is considered to be 'a process that takes away the ability of individuals and their communities to fully function, through poor health, destruction of economic livelihoods, and general widespread environmental threats' (Schlosberg 2007: 80). It follows from these conceptions of environmental justice that the consequences of climate change make it a case of widespread and frequently extreme injustice. Indeed, it is a matter of *global* environmental justice because it involves 'the global distribution of environmental burdens and benefits' (Caney 2005a: 748).

Climate justice can be defined very generally as the 'special problems of obligation and participation posed by climate impacts and policies for their management' (Beckman and Page 2008: 527). Ludvig Beckman and Edward Page (2008: 528) point to the important question of who should be the 'recipients' of climate justice, that is, 'the question of which entities (individuals, groups, countries, generations) possess claims against others that their atmospheric security be respected', which in turn requires that we identify 'the entities that bear the "burden" of guaranteeing that distributive entitlements are respected'. As we will see in Chapter 4, if one answers these queries by looking at the existing international agreements on climate change, the answer is, almost always, that *countries* (that is, states) are the entities that have been officially identified as both the recipients of climate justice and the bearers of associated burdens.

As a prelude to more detailed arguments in subsequent chapters, we can begin with what might be called a 'generic argument' for 'extensive duties of climate justice' (Beckman and Page 2008: 528): '(1) acts or policies that modify the atmosphere threaten the interests of future persons; (2) human activities that threaten the interests of future persons are unjust; therefore (3) acts or policies that modify the atmosphere are unjust.' It follows from this that past acts or policies threatening the interests of persons today were, at least potentially, unjust when they occurred, or at least that they might impose some duties on those who are now benefiting from them. Justice obtains when we cause harm to others; this causal connection demands action. As Bill McKibben (1989: 46) points out, climate change will impact 'every inch and every hour of the globe'. Thus Dobson (2006: 175) suggests that climate change offers perhaps the most concrete example of 'global relations of causal responsibility' that invoke requirements of justice. As Lorraine Elliott (2006: 345) declares, 'global politics of the environment is increasingly a politics of transnational harm that raises important questions about injustice and global ethics'. Mark Smith and Piya Pangsapa (2008: 1) believe that 'environmental issues cannot be separated from questions of social justice'. Caney (2008: 537) argues that people 'have a right to a healthy environment', 'persons have a right not to suffer from dangerous climate change', and the 'current consumption of fossil fuels is . . . unjust because it undermines certain key rights'. For Stephen Gardiner (2004: 578–9), insofar as climate change is a matter of morality, it is primarily one of justice: 'The core ethical issue concerning global warming is that of how to allocate the costs and benefits of greenhouse gas emissions and abatement'. For Singer (2006: 415), climate change is an ethical issue because of the associated injustices: 'Climate change is an ethical issue because it involves the distribution of a scarce resource – the capacity of the atmosphere to absorb our waste gases without producing consequences that no one wants'.

Principles of climate justice, specifically concepts of responsibility across borders (and generations), might be placed into one of three categories (Jagers and Duus-Otterstrom 2008: 578–9): polluter pays ('perpetrators of some environmental damage ought to bear the costs of its bad effects' (Page 2006: 53)), ability to pay ('parties who have the most resources should contribute the most' (Shue 1999: 537)) and a hybrid of the two proposed by Caney (2005a), who argues for a distributive theory 'which attaches weight both to wealth and [to] causal contribution' (Jagers and Duus-Otterstrom 2008: 584). One advantage of the latter approaches is that at least they allow us to say that well-off states or people (and other actors) ought to take action *now* to mitigate their emissions and to help people adapt to climate change. The question of who is responsible for past emissions is vexing indeed, but it is most vexing because it prevents action. Any principle that credibly and acceptably promotes action by advantaged people living right now deserves attention.

Climate change is only the most recent and most profound manifestation of a general rule of thumb: the world's natural wealth is distributed very unevenly, with about 25 per cent of the people on the planet appropriating about 75 per cent its resources (Sachs and Santarius 2007: 135). As Elliott (2006: 348) points out, those people 'most immediately affected by global environmental decline are those who have contributed least (or certainly proportionately less) to the problem'. The greenhouse gas emissions of mostly affluent countries and people will result in extreme harm and suffering among poor countries and people. Thus Vanderheiden (2008: 45–6) argues that 'climate change presents a case of the world's affluent benefiting at the expense of the world's poor in a relationship that can plausibly be described as exploitation'. Consequently, climate change is a matter of justice or, rather, injustice committed by the world's affluent people. Very importantly, most people who participated in this extensive consumption of natural resources, including consumption (that is, pollution) of what was an otherwise pristine global atmosphere, were until quite recently mostly in the developed states. However, nowadays a growing proportion of this group of consumers comprises affluent 'new consumers' living in developing countries such as Brazil, China and India (Myers and Kent 2004).

CONCEPTUALISING THE INJUSTICES OF CLIMATE CHANGE

Climate change is a profound matter of injustice imposed on people who are already poorly off and who usually have no say in the matter. The adverse effects of climate change most harm the weakest and poorest countries and people of the world, imposing burdens on those states and those people least responsible for causing it, most exposed to it and most vulnerable to its ravages (for example, sea-level rise and storm surges, drought and floods, disease and heat stress) and least able to pay for mitigation and adaptation. These effects are mostly a result of what other, mostly affluent, people have done in places far from where the pain is felt. Consequently, because of its causes and its consequences, climate change is an issue that cries out for justice.

Looking at several accounts of justice can help to illustrate and reinforce this assertion that climate change is indeed a matter of justice – and injustice. Each account may take us in different directions, but overall they provide a powerful set of moral arguments for seeing climate change in terms of justice and for seeking ways to overcome the associated injustices. Here I will discuss several widely accepted (but hardly exhaustive) accounts of justice: causality and responsibility, utilitarianism, Kantianism, Rawlsian philosophy, impartiality and basic rights (Brown 1992: 159–92; cf. Paterson 1994, 1996; Harris 1996).[1]

CAUSALITY AND RESPONSIBILITY

According to conceptions of justice based on causality and responsibility, those not responsible for causing a problem (for example, pollution) should not have to pay to fix it, and those responsible for causing harm are at least responsible for righting it.[2] Henry Shue (1995: 386) argues that 'the obligation to restore those whom one has harmed is acknowledged even by those who reject any general obligation to help strangers . . . One virtually always ought to "make whole", insofar as possible, anyone whom one has harmed . . . because one ought even more fundamentally to do no harm in the first place.' Put another way, 'if it is the case that the poverty [or, we could add, suffering from adverse environmental change] of poor countries/peoples is the result of actions by rich countries/peoples, then there would seem to be quite a strong *prima facie* case for saying that the latter have a clear responsibility to act in such a way as to make reparations . . .' (Brown 1992: 159). Thomas Pogge (2005: 46) points out that 'our moral duty to help protect people from harm in whose production we are implicated [is] more stringent than our moral duty to help protect people from harm in whose production we are not materially involved'.

If we think in terms of states, there is no disputing that the developed industrialised countries and their citizens are inordinately responsible for the historical pollution causing global warming and the resulting climate change. With only 20 per cent of the world's population, the developed countries have produced three-quarters of the world's historical carbon emissions, benefiting economically as a consequence (Speth 2008: 28). The United States, with one-twentieth of the world's population, alone produces close to one-quarter of the world's greenhouse gases, and per-capita emissions of greenhouse gases in the United States in the early twenty-first century are roughly ten times the global average. Energy use per capita in the industrialised countries is over thirty times that in all developing countries – and one hundred times that in the least developed countries. If we accept that the affluent countries are indeed responsible for a disproportionate share of global pollution, then the debate can move to questions of what to do about it. The so-called polluter-pays principle is an accepted standard within affluent countries.[3] Interpretations of international justice based on causality and responsibility suggest that this principle ought to be applied among national communities, not just within them.[4] However, the way that the climate change regime has implanted the principle 'can be characterized as an instance of the polluted pay principle: polluters benefit from their pollution, while those who suffer from the pollution bear the costs' (Jamieson 2005: 168).

Determining responsibility for environmental pollution is often very difficult, especially when we are addressing issues as complex as climate change. Even if we agree that the affluent countries deserve the bulk of the blame for the problem, assessing which of them have caused how much harm to which poor (or other affluent) countries is a daunting task indeed, especially in the light of rapidly increasing emissions of greenhouse gases from the poor countries.[5] Nevertheless, what this interpretation of international justice calls for is, at least, that the world's poor countries 'ought not be asked to sacrifice in any way the pace or extent of their own economic development in order to help to prevent the climate changes set in motion by the process of industrialization that has enriched others' (Shue 1992: 395). Vanderheiden (2008: 81) is persuasive when he reminds us that those states and people who 'bear the brunt of climate-related damage are among those least responsible for the GHG [greenhouse gas] pollution causing climatic problems, and those that are the most responsible for current and historical GHG emissions are expected to suffer the least damage'. What seems beyond much dispute is that, from the perspective of justice based on causality and responsibility, the ills caused by some and suffered by others make climate change a matter of (in)justice.[6]

UTILITARIANISM

A utilitarian might argue that the distribution of resources should be justified based on the total amount of happiness (or 'utility') it produces, measured by summing up the happiness experienced by individuals. The utilitarian would probably look at the consequences for promoting overall happiness of distributions based on rights, desert, need and so forth (D. Miller 1976: 32). Anthony Ellis (1992: 164) suggests that classical utilitarian philosophers, such as Jeremy Bentham, were fundamentally cosmopolitan, believing that people are simultaneously citizens of their own nations and of the world, with duties to the good of humankind in general. Bentham (1962: 538) wrote that it is criminal for a country 'to refuse to render positive services to a foreign nation, when rendering of them would produce more good to the last mentioned nation, then it would produce evil to itself'.

A potential problem with utilitarianism is that its goal of achieving a distribution leading to the greatest overall utility could result in some individuals suffering if such an outcome contributes to overall, but not always individual, happiness (Almond 1995: 12). This, on the face of it, seems to be unjust. But, in the context of the severe effects of climate change, it may be fair to have the few 'suffer' fewer luxuries to save the planet on which all people depend for health and survival. It hardly seems just to let people exercise some individual 'rights', such as the right to spew carbon dioxide while driving sporty cars for the thrill of it, if doing so will contribute to the adverse effects of climate change described above. As Shue (1992: 397) puts it, 'whatever justice may positively require, it does not permit that poor nations be told to sell *their* blankets in order that rich nations may keep *their* jewellery'. Singer (1996) argues that, if it is in our power to prevent something very bad from happening, without thereby sacrificing anything that is morally significant, we have a duty to do it. Furthermore, because we are able to do so, we ought to act to prevent, for example, the starvation of thousands of people outside our society, even if this means sacrificing (for example) the upholding of property norms in our own society. Singer argues that most people in the affluent countries ought to go to great lengths to help those abroad, certainly doing much more than they are at present to provide this help. Indeed, he argues that 'we ought, morally, to be working full time to relieve great suffering of the sort that occurs as a result of famine or other disasters' (Singer 1996: 33), which would, of course, include these effects resulting from climate change. To use Singer's illustration (1996: 36) regarding famine relief, 'it should be clear that we would have to give away enough to ensure that the consumer society, dependent as it is on people spending on trivia rather than giving to famine relief, would slow down and disappear entirely'. This position may seem a bit extreme, but surely trading 'trivia' for protection of the Earth's atmosphere would be a reasonable trade-off.

From the utilitarian perspective we might say that individuals, with the assistance and encouragement of governments, ought to take action to promote what could be called the utility of environmentally sustainable development. This could mean that the affluent countries and people would aid poor countries and people to develop and live sustainably because to do so would simultaneously reduce human suffering (and thereby increase overall 'utility') and reduce – and potentially reverse – environmental destruction, which could otherwise minimise happiness in the future. Utilitarianism compels us to look at overall consequences of our individual actions and suggests that it may be appropriate to treat some individuals differently if the environment vital to all life will be protected as a consequence. From this perspective of justice, national boundaries are (conceptually) of little relevance; the goal is global rather than local (or national) utility maximisation. This requires one to find a measure of utility that can be used to allocate benefits and burdens – a difficult task indeed, at least regarding maximising the utility of costs and benefits associated with addressing complex environmental issues like climate change. But this difficulty should not preclude us from thinking in these terms and trying to maximise utility defined in terms of environmentally sustainable development. More generally, viewing climate change from a utilitarian perspective shows, again, that it is a matter of justice.

KANTIANISM

According to a Kantian conception of ethics, all humans have obligations towards one another by virtue of their common humanity. Immanuel Kant's 'categorical imperative' requires that people be treated as ends in themselves, rather than as means to someone else's ends (Kant [1785] 1948).[7] According to O'Neill (1996: 97), 'we use others as *mere means* if what we do reflects some maxim *to which they could not in principle consent*'. From this perspective we might say that it is unjust to exploit other persons or to deny their human rights because to do otherwise would not receive their consent. Kantian principles of human equality and right and wrong action might be used to assess obligations of people in different countries towards one another. Categorical imperatives could be established for world affairs. Duties of equity, by the Kantian conception, exist whenever there is involvement between actors, something that is common and widespread in the modern world (O'Neill 1996: 103). O'Neill (2000) believes that the common humanity shared by all people requires that we meet basic minimal obligations towards one another, with these obligations deriving from basic Kantian assumptions, such as truth telling and non-coercion in social relations. If we are all to treat one another as ends, we must each have the capacity to do so: 'our obligations as moral agents require us to help others to be moral agents' (Brown 1992: 170). It may follow that affluent governments and people ought to start by assisting poor people in being moral agents (O'Neill 1986). Thus in a discussion of ending world hunger, O'Neill (1996: 99) declares that 'the requirement of treating others as ends in themselves demands that Kantians standardly act to support the possibility of autonomous action where it is most vulnerable [and] to do what they can to avert, reduce, and remedy hunger. They cannot of course do everything to avert hunger: but they may not do nothing.'

Extending this argument to climate change, citizens of all countries might be obligated to refrain from unsustainable use of natural resources or from pollution of environmental commons shared by people living in other countries – or at least be obligated to make a good effort towards that end. (Poverty alleviation would be necessary before many people could fulfil such obligations in poor countries.) Insofar as the actions of individuals, corporations and governments contribute to the effects of climate change, if those effects are imposed on people or communities without their consent, or if they limit others' abilities to act as moral agents, we can say that the actions are a violation of Kantian justice. Affluent countries and people need not return to the Stone Age when fulfilling these obligations. Rather, what this conception seems to demand is that these considerations be taken seriously in decision making. Kantian ethics aim to 'offer a pattern of reasoning by which we can identify whether *proposed action or institutional arrangements* would be just or unjust, beneficent or lacking in beneficence . . .' (O'Neill 1996: 103).

Some observers argue in favour of the duty to ensure freedom: 'The point of justice is not to guarantee a good life for every citizen of the world, but rather to leave everyone free to follow their own project for a good life . . . Everyone does have a common interest in the freedom to live in their own way and by their own lights. A duty-centred theory of justice [following Kant] will ground the injunction of responsible behaviour on the demand to respect the freedom of others' (Sachs and Santarius 2007: 137). A scheme for a just distribution of the world's resources would follow rules that could be universalisable, and 'over-appropriation of the environment by a few strong countries at the expense of many weaker ones contradicts such rules' (Sachs and Santarius 2007: 139). Over-appropriation of resources in this way would effectively rule out the opportunity for many people to enjoy a good life (and could rule out life itself). Thus seriously considering Kantian conceptions of justice in the context of climate change, and trying to find ways to make actions and policies fit these conceptions, might go a long way towards promoting Kantian justice. At the very least, we can conclude once again, but from another viewpoint, that climate change is an important matter for anyone concerned about justice.

RAWLSIAN JUSTICE

John Rawls (1971) developed a contractual theory of social justice (at least partly in response to the deficiencies he detected in utilitarianism) comprising some fundamental principles of justice: the equal-liberty, difference and fair-opportunity principles. The latter two can be especially useful for thinking about justice and climate change. According to the difference and fair-opportunity principles, social and economic inequalities should be arranged so that they are both 'to the greatest benefit of the least advantaged' (the difference principle) and 'attached to offices and positions open to all under conditions of fair equality of opportunity' (the fair-opportunity principle) (Rawls 1971: 303). Inequalities in distribution are acceptable insofar as they benefit the least advantaged in society, and they should be arrived at based on equality of opportunity. Rawls justifies these principles by saying that they fit our intuitive moral judgements (they are 'common sense') and rational individuals would agree on principles of a just society if they were to make their choice in the 'original position' behind a 'veil of ignorance', where they would not know their nationality, position in society, skin colour, religion and so forth.

Charles Beitz (1979b) applies Rawls's theory to international relations based on the observation that international society is sufficiently cooperative and interdependent to be classified as a 'community'.[8] According to Beitz (1979b: 151), global interdependence has created a situation in which national societies are neither self-contained nor self-sufficient, and

> if evidence of global economic and political interdependence shows the existence of a global scheme of social cooperation, we should not view national boundaries as having fundamental moral significance. Since boundaries are not coextensive with the scope of social cooperation, they do not mark the limits of social obligations. Thus the parties to the original position cannot be assumed to know that they are members of a particular national society, choosing principles of justice primarily for that society. The veil of ignorance must extend to all matters of national citizenship, and the principles chosen will therefore apply globally.[9]

Rawlsian principles can help to illustrate the justice implications of climate change. Like other aspects of globalisation, climate change is driving countries and people together in cooperative efforts to prevent mutually destructive consequences, and it is the least advantaged in world society – the least developed countries and poorest people of Africa, Asia and Oceania – that are most routinely vulnerable to the impacts of climate change mostly caused by activities in more affluent places. In those instances where the environmental plight of weak countries and peoples – such as the small-island states and their citizens, which are particularly vulnerable to the effects of climate change (for example, sea-level rise) – are being taken seriously by the international community, we might say that Rawls's concern for the least advantaged have been recognised, or at least there is a move in that direction. Furthermore, the equal-opportunity principle that is part of Rawls's theory of justice, if applied to international relations, suggests strongly that diplomats should give serious consideration to equal opportunity in cooperative arrangements – international institutions – intended to address climate change. This would mean giving poor country governments and poor people more of a say in decisions regarding the sharing among countries and peoples of benefits and burdens associated with climate change. Indeed, this is what has happened in several international environmental financing institutions, such as the United Nations' Global Environment Facility, the Montreal Protocol's Multilateral Fund and the Adaptation Fund of the climate change convention (see Chapters 3–4). Thus, the application of Rawlsian principles to climate change can further establish the issue as one of (in)justice.

IMPARTIALITY

According to a conception of justice described by Brian Barry (1989, 1995; see also Brown 1992: 179–82), actors have interests that sometimes come into conflict, but they use the 'arsenal of persuasion' to arrive at agreements on terms that no one of them could reasonably reject. The motive for behaving justly is, according to Barry (1989: 8), 'the desire to act in accordance with principles that could not reasonably be rejected by people seeking agreement with others under conditions free from morally irrelevant bargaining advantages and disadvantages . . . The significance of speaking of "justice as impartiality" is that [people should] seek to find a basis of agreement that is acceptable from all points of view.'[10] Barry (1989: 8) shows that it is difficult to persuade actors to negotiate based on self-interest and reciprocity if those actors do not wish to reach a reasonable agreement. The question becomes: what is reasonable? Mutual advantage as the underlying goal of cooperation is replaced by a desire to reach agreement. Barry envisons substantial redistributions between countries to ensure that each has a fair share of the world's resources. Countries would agree to such transfers because they are reasonable, not because they expect a reciprocal response. As Brown (1992: 181) describes this viewpoint, 'as between the United States and Bangladesh, there can be no reciprocity – for the foreseeable future the relationship will be one-way. The United States should aid Bangladesh not because it is in the United States' interest to do so but because justice as impartiality suggests that the case for such aid cannot be reasonably denied.'

Countries and people should ask themselves what is reasonable to expect of one another. Is it reasonable to continue to emit pollutants that contribute to climate change that will have especially adverse effects for the poorest countries and people, and is it reasonable to deny poor countries and people the help they require to cope with climate change and to join in efforts to prevent it, especially when the developed world is disproportionately at fault? Barry's argument would justify greater egalitarianism between countries, including transfers between rich and poor countries and people to address climate change. Furthermore, Barry (1999: 37) argues that 'the value of any political structure (including a world state) is entirely derivative from whatever it contributes to the advancement of human rights, human well-being, and the like'. While he concedes that cosmopolitan morality does not require a world state, he points out that 'the present system is incapable of dealing with such vital issues as global warming, the loss of biodiversity, and pollution' (Barry 1999: 39). Matthew Paterson (1994: 19) suggests that Barry's framework offers the most convincing grounding for justice in the context of climate change. Barry's perspective points not only to the justice implications of climate change, but also to the particular kind of justice advocated in Part III.

BASIC RIGHTS

There is now a well-established tradition of trying to apply conceptions of human rights to international affairs, although until the mid-twentieth century rights were most often assumed to be held almost exclusively by states (see Chapter 3). From the perspective of human rights, one might say that individuals have inherent rights to minimum nutrition, freedom from torture, freedom of expression and so forth, simply because they are human beings. At the very least, we might say that individual persons ought to have their security and subsistence rights (however defined) respected, for without those rights all other rights cannot be fulfilled (Shue 1996a). Rights can be difficult to apply in practice to world affairs because their establishment often begs the questions of who or what (for example, individuals, peoples, governments) is entitled to which rights and who or what is responsible for fulfilling them. And, while human rights are now very much on the agenda of states, so far little has been done to promote the right to development or to a clean environment.[11]

Nevertheless, notions of human rights help identify some of the ways in which climate change is a matter of justice. Shue's (1996a) principle of 'basic rights' is especially instructive. In trying to show that something ought to be a right, Shue (1996b: 114) argues that we need to show that it is vital – either literally necessary to, or highly valuable to, living as a human – and that it is vulnerable – subject to widespread threats that individuals on their own often could not defend against. At the very least, humans have a right to survival, for without life all other rights cannot be exercised. David Held (1995: 195) argues that all people have 'a right to a clean, non-toxic, sustainable environment' because denying this right would prevent people from adequately participating in political life. Shue (1996a) says that there are duties attached to human rights: basic human rights require actors to avoid depriving, to protect from deprivation, and to aid the deprived. Depending on the circumstances, these duties may be ascribed to different actors (for example, individuals, non-governmental organisations, state governments), although states, by virtue of their capacities, may be the most important duty-bearers, and the nature of the duty may vary – for example, to avoid depriving individuals of the right to subsist, to protect refugees from assault, or to bring aid in times of natural disaster.

Simon Caney (2008: 551) argues that climate change is unjust because it violates human rights: 'Climate change undermines persons' human rights to a decent standard of health, to economic necessities, and to subsistence.' He points out that all people have 'fundamental interests in health, subsistence and supporting themselves', everyone has the 'right not to suffer from climate change which jeopardises these interests' and these interests are 'sufficient to impose duties on others', not least because acting on these duties would not be 'unreasonably demanding' (Caney 2008: 539). However, in the climate change negotiations,

emission thresholds are fixed which keep to a minimum the losses to industrial countries but happily accept the loss of survival rights among fishing communities, farmers and delta inhabitants in the southern hemisphere. It is contrary to justice when some gain advantages for themselves at the cost of great disadvantages for others. Fairness demands shelving advantages to oneself if they would further harm those who are already weak. The minimal rule of distributive justice thus coincides with the respect for existential rights throughout the world. It is anyway unjust to sacrifice the survival needs of some to the prosperity needs of others. (Sachs and Santarius 2007: 137)

The injustice of climate change identified by this viewpoint has been shown by the Intergovernmental Panel on Climate Change (IPCC 2002: 12), which reported in 2001 that the impacts of climate change 'will fall disproportionately upon developing countries and the poor persons within all countries, and thereby exacerbate inequities in health status and access to adequate food, clean water, and other resources'. Perhaps with these kinds of considerations in mind, the United Nations Human Rights Council recognised in 2008 that climate change has implications for human rights. The council expressed concern that 'climate change poses an immediate and far-reaching threat to people and communities around the world and has implications for the full enjoyment of human rights'; recognised that 'human beings are at the centre of concerns for sustainable development and that the right to development must be fulfilled so as to equitably meet the development and environmental needs of present and future generations' and that 'the world's poor are especially vulnerable to the effects of climate change, in particular those concentrated in high-risk areas, and [the poor] also tend to have more limited adaptation capacities' (Human Rights Council 2008: 65).

Insofar as climate change denies people or entire communities the capacity to survive, it would follow from the basic-rights perspective that individuals who consume and pollute more than necessary have some obligation to stop doing so. It also follows that governments should actively try to stop emissions of pollutants from within their jurisdictions that harm people in others. Shue's argument might also support an obligation actively to try to help people face the challenges of climate change, to help them implement their own sustainable development measures and to help them cope with the consequences of, say, sea-level rise, such as in the case of the small-island states and their residents. To be sure, it is difficult to attach specific obligations to the denial of subsistence resulting from climate change – who caused how much greenhouse gas pollution taking away how much of which person's or which group's ability to survive? – but just because it is difficult to identify the details of how to operationalise rights in this regard does not mean that injustices should be ignored. In specific cases it may be possible to identify who is harming other classes or groups of people, and which activities the former ought to stop and what they should do to help those they have harmed (or will harm). It is important to put polluters on notice that they are probably violating some person's or some country's rights when adding unnecessarily to global warming. At least we can say that, from the perspective of human rights, many of the causes and especially the impacts of climate change make it a matter of (in)justice.

CONCLUSION

The characteristics of climate change make it a matter of justice. For example, the way in which climate change connects people together through causes in one place impacting on people in another requires one to consider the justice implications. In the context of climate change, the fate of people everywhere is connected; we all share the same environmental space. Environmentalists might say that this goes without saying. The famous photograph of the entire Earth taken by Apollo astronauts during their journey to the moon made this clear. But it has been less obvious to communitarians. Elliott (2006: 346) argues that environmental change is a matter of justice because the harm and suffering that come from it 'are unevenly distributed and (this is the important feature) that unevenness is unjust, not simply because some are more likely to cause it and others are more likely to suffer it but because the causing and the suffering are increasingly linked in a complex web of responsibility and displacement'. This displacement of environmental harm is inequitable in three ways. First, 'globally the rich consume more resources and produce more waste than the poor' (Elliott 2006: 348). Second, there are winners and losers, with those causing the most environmental harm often winning while the least responsible lose the most. Third, those most affected by environmental harms are the least likely to be involved in the making of decisions about how to deal with pollution and the use of environmental resources.

As Wolfgang Sachs and Tilman Santarius point out, not only are global pollution and resources distributed unevenly among rich and poor countries, but so too is the damage that results, and with 'all probability, the bitter effects of climate change will hit first and most powerfully the countries and people who did least to cause it' (Sachs and Santarius 2007: 53). This unequal distribution of the causes of climate change makes it a question of justice, and the fact that people not responsible for climate change will suffer its consequences also makes it a matter of injustice. These consequences are inevitable. Consequently, 'the cardinal climate change inequity is . . . not the *potentially* unfair allocation of mitigation targets [which have preoccupied the climate change regime] but the *inevitably* unfair distribution of climate impact burdens' (Muller 2002: 4). As Vanderheiden (2008: p. xiii) describes the situation, it is the world's poorest countries and peoples who will bear the brunt of the impacts of climate change largely caused by the world's affluent people, meaning that it 'presents a unique case of global injustice, where the ongoing failure to adequately address the problem exacerbates the global inequity that is part and parcel of the problem itself'.

Many accounts or views of justice can be brought to bear to show that climate change is a matter of profound injustice, but surely the causes (by the world's affluent) and the consequences (for the world's poor) are among the clearest reasons for saying so. As we will see in Chapter 4, governments have come to recognise that climate change is a matter of justice, although the way that this recognition has been interpreted and manifested means that justice has seldom been realised. But, before looking at justice in the climate change regime, Chapter 3 will examine how the idea of justice has found its way into international environmental affairs more generally.

NOTES

1. In Brown's terminology (1992), the following accounts are varieties of international justice. Cf. Dower's (1997: 566–7) sources of cosmopolitanism justice. This section is developed from Harris (2001a).
2. From this perspective we might say that someone responsible for causing harm ought to correct it but also go further by, for example, paying additional compensation to those harmed, paying a fine or otherwise being punished, the latter intended in part as a negative incentive or as a message to others that could deter future harm.
3. Member countries of the Organisation for Economic Cooperation and Development agreed in 1972 to base their environmental policies on the polluter-pays principle.
4. Having said this, the polluter-pays principle does not fully capture the moral significance of all historical causal responsibility. Thus Muller (2002: 32) prefers the German expression Verursacherprinzip – the 'principle of (who or what is) the cause'.
5. One especially difficult question surrounds assessing responsibility for historical emissions of greenhouse gases because the developed countries and their citizens did not know they were harming other countries until quite recently. This raises the paradoxical notion that the developing countries, and many (affluent) people living there, may have a higher standard of duty than did the developed world at the same stage of industrialisation (see Chapter 6).
6. One might argue that causality does not constitute a moral theory on the same level as the others discussed below, and that it could be seen as a presupposition of them. Even if the former is true, the latter may not always be so. As we shall see, climate justice may obtain even when there is no causality. For example, affluence and capability may fully justify certain obligations; we may be obligated to limit our contributions to climate change simply because we can – although, if those contributions harm others, the obligation is arguably much greater.
7. This 'formula of the end in itself' is but one of Kant's formulations of his categorical imperative (see O'Neill 1993).

8. Rawls says that if countries were to come together in an original position they would choose principles of justice familiar to international lawyers: equal rights of states, self-determination and non-intervention, right to self-defence, justice of and in war (jus ad bellum, jus in bello) and the like (Rawls 1971: 377–82 and passim). In his *The Law of Peoples* (Rawls 1999) he argues for international justice in the form of commitments to a constrained set of human rights and protections for sovereign peoples honouring those rights (see Rawls [1993] 2008), rejecting arguments by Beitz and others to extend his theory to global distributive justice. Some scholars argue that Rawls was trying to offer an account of 'global justice' (Brock 2009: 19–40), although not in the sense used here.

9. Beitz subsequently changed his reasoning. While he thinks that Rawls's difference principle still ought to apply to international relations, he bases this conclusion on a Kantian perspective (see Beitz 1983, 1991).

10. This is analogous to Kantian conceptions of justice, as Barry acknowledges.

11. The right to development was codified in the 1986 United Nations Declaration on the Right to Development. Dower (1992) suggests that there should also be a right to environmentally sustainable development.

PART II
INTERNATIONAL JUSTICE

CHAPTER 3

INTERNATIONAL ENVIRONMENTAL JUSTICE

For centuries the world has been guided by, and governments have sought to reinforce, norms of state recognition, sovereignty and non-intervention. According to these prevailing and powerful norms, sovereign states are the legitimate expressions of human organisation, and it is to states that people ought to turn for governance and for solutions to major challenges. These norms have so far largely guided discourse, thinking and responses to transboundary environmental problems: environmental diplomacy, regimes and treaties have been based (by definition) on the responsibilities, obligations and capabilities of *states* to limit pollution affecting other states, to share resources in areas not controlled by individual states, and generally to cooperate to cope with the effects of environmental harm and resource exploitation. Indeed, the international norms have been so powerful as to result in what is, effectively, a doctrine of *international* (or, more precisely, inter-*state*) environmental justice, as described in this chapter, and manifested in the principle of common-but-differentiated responsibility among states, which is examined in the next chapter in the context of climate change. This international doctrine has guided and permeated a number of international environmental agreements and regimes, such as the treaties to combat stratospheric ozone depletion and manage biological diversity. While it is possible to identify some beneficial consequences of these responses, in the case of climate change Westphalian international norms have acted as a kind of curse, stifling diplomatic and human creativity – and action – by fostering a clash between rich and poor states that largely ignores the rights and duties, as well as the suffering, of *individuals*.

INTERNATIONAL JUSTICE AND THE STATE

Most people take the interstate system for granted. As introduced in Chapter 2, and as we will see in more detail in Chapter 5, a major exception can be found among cosmopolitans. The world was not always comprised of sovereign states as we recognise them today. Rather, the existence of territorially defined, self-governing modern states is a relatively new phenomenon, dating from around the middle of the seventeenth century. It was very recently, during the decades following the Second World War, that the idea of the sovereign state became almost immutable in people's minds, with membership of the United Nations ballooning from about 50 states when it was founded in 1945 to a number approaching 200 today. Somewhat paradoxically, this expansion in the number of states occurred just as globalisation, which is often defined as the *erosion* of territorial boundaries and state sovereignty, began in the 1970s and became wider, faster and deeper in the final decades of the twentieth century. Yet, despite economic and other forces that are undermining the abilities of state governments to control what happens within their own borders, the idea of territorially defined political communities governing their own affairs 'remains central to the orthodox state-centric conceptions of world politics as the pursuit of power and interests between sovereign states' (McGrew 2008: 23). Not surprisingly, environmental regimes have been shaped by this state-centric orthodoxy – this international doctrine.

THE STATE SYSTEM

The international system we take for granted evolved through a gradual process over hundreds of years, but it is generally dated to the Treaty of Westphalia of 1648 (and other agreements from around the same period), which codified the fundamental principles and norms that still guide – or at least substantially and often materially influence – world affairs today. The feudal entities that preceded Westphalia were characterised by a lack of monopolistic political authority. Local rulers instead routinely paid homage to a number of higher powers, as occurred in the Middle Ages when loyalty was given simultaneously to local and regional feudal rulers, the Holy Roman Emperor and the Pope. During much of European history, politics was characterised by relationships among nobles, religions, cultures and the like, rather than among territorial states. This changed as monarchs sought to consolidate their authority through war, which was exemplified in the expanding economic and political footprints of cities. The wars that were necessary to garner territory and wealth required technological, economic and bureaucratic innovations. European cities obtained greater resources through expansion, order through the development of centralised bureaucracies, and security through military technology, which in turn required more wealth and greater political consolidation of expanding territories obtained through war. This economic and political expansion, coupled with the bureaucratic developments that enhanced the ability of rulers to control larger territories and to extract wealth from them, allowed rulers to centralise their power over discrete territorial spaces. Thus, as states made war, war made states (Russett and Starr 2004: 47–9; cf. Tilly 1975).

The consolidation of modern-style states came about when secular monarchs challenged the spiritual authority of the Catholic Pope in Rome, in the process sparking the Thirty Years War among European monarchies. These wars ended with the Treaty of Westphalia, which was premised on a new principle: that the ruler of a particular territory, not the Pope or another external power, would decide the religion practised there. Consequently, the basis for the sovereign state, with ultimate authority being held by each state's ruler rather than by an external power, was established (unlike in other parts of the world, where suzerainty prevailed). Rulers thus became sovereign, enjoying the exclusive right of authority over a particular piece of territory. Territory, not allegiance to another ruler or religion, became the basis of authority (Russett and Starr 2004: 47–9).

Thus the Treaty of Westphalia was a primary legal and normative foundation for the modern state and the state system that took shape in subsequent centuries, epitomised in today's international laws, norms and practices. The fundamental principles of the Westphalian international system that have emerged since 1648 are territoriality, sovereignty and non-interference. The world is now organised into territorially exclusive states where each government has ultimate legal authority within its borders, in principle free from interference by other states. To be sure, many weak states have not been able to exercise all these principles. Nevertheless, the Westphalian norms remain powerful in shaping the fundamentals of world affairs in most issue areas. Nigel Dower (2007) identifies a number of key elements of the Westphalian 'morality of states' that form the basis of the international system, of state behaviour (most of the time), of international agreements generally and indeed of the climate change regime. Among these elements are the duty of states to support the state system; to 'respect the sovereignty, autonomy and independence of other states'; not to 'interfere or intervene in the internal affairs of other states'; and not to 'harm other states, but [there is] no duty in general to promote the global common good, the good of other states or the good of individuals living in other states', the latter because 'individual human beings do not have rights against any states other than those they live in' (Dower 2007: 59). This final point is important for our understanding of the climate change agreements and other international environmental agreements.

The principle of non-interference in the internal affairs of states was codified in, and indeed permeates, the Charter of the United Nations. The Declaration on the Inadmissibility of Intervention and Interference in the Internal Affairs of States, adopted by the General Assembly in 1981, affirms the fundamental principle of 'non-intervention and non-interference in the internal and external affairs' of states, declaring 'the sovereign and inalienable right of a State freely to determine its own political, economic, cultural and social systems, to develop its international relations and to exercise permanent sovereignty over its natural resources, in accordance with the will of its people, without outside intervention, interference, subversion, coercion or threat in any form whatsoever' (United Nations 1981: para. I(a)). This declaration is indicative of states' responses to environmental problems. As we have seen already, however, while states may respect the principle of non-interference in what happens within other states, pollution does not.

INTERNATIONAL JUSTICE

As we saw in Chapter 2, justice has a place in the relations among sovereign states. The notion of justice in international relations is informed by prototypical communitarian conceptions of justice within domestic societies. According to John Rawls (1971: 4), a society is defined as a 'cooperative venture for mutual advantage'. Individuals 'pre-exist' in society and cooperate to produce more goods, however defined, than would accrue from non-cooperative behaviour. According to the communitarian world view, people develop their identities and measure value based on the communities, routinely the states, in which they live. Definitions of justice stem from particular communities and thus apply within them. From this perspective, states have very few duties of justice towards one another, and even less so towards people living in other states. Consequently, a communitarian might argue, we should not apply principles of justice to international relations.

But not everyone would agree with such restrictions, and in practice principles of justice do apply rather widely to international affairs. States have for centuries accepted (often in the breach) principles of 'just war', and since the mid-twentieth century principles of distributive (or 'social') justice have become increasingly prominent in relations among states as the more affluent among them have helped the less affluent, and as states have increasingly recognised some routine duty in doing so. In recent decades there has even been a gradual acceptance among many states that there are duties of justice between well-off states and *people* in other states who are suffering from starvation, widespread human-rights abuses and persecution. This suggests that the most restrictive conception of the 'morality of states' is changing.

Simply put, justice calls for the differences that are created by social arrangements to be justified. At the international level, differences between states and people are often the result of international interaction and cooperation. It would follow, then, that justice requires that these international differences be justified somehow. Conceptions of international justice, according to Chris Brown (1992: 171), 'ought to allow us to place relations between rich and poor countries on a footing that recognizes diversity yet meets the obligations people have towards one another – or, at least, tries to perform the same sort of role that notions of justice try to perform in domestic society'. Justice is particularly salient in international relations in the context of economic globalisation and environmental interdependence. The twentieth century witnessed increasing disparities in wealth between North and South (and within many national communities), and these disparities look set to remain or even widen in coming years. Many people would argue that these differences are a consequence of actions by economically powerful states. Increasing awareness of these disparities made possible by modern communications and information technologies, travel and trade, and – very importantly – the increased capacity to redistribute resources around the globe, have given prominence to issues of international distributive justice. It is not possible to deny that hundreds of millions of people in the poor countries are suffering, nor is it possible to deny that the developed countries are able substantially to reduce this suffering. More specifically, it is evident that actions in the developed countries are causing harm to the developing countries. For example, pollution of shared environmental commons and resources in the former causes harm to large numbers of people in the latter. Much or all of this harm is arguably unjust and certainly avoidable.

International justice has been on the agenda of international politics for decades. Demands by poor countries for more just treatment in international economic affairs were evident when the United Nations Charter was being negotiated. This was reflected, for example, in the placement of the Economic and Social Council alongside the other principal organs of the United Nations (Luard 1994: 160). Relations between rich and poor countries came onto the international agenda with much greater prominence in the 1960s as a result of post-Second World War decolonisation and as the developing countries acquired a majority of votes in the United Nations General Assembly, putting them in a stronger position to push their economic demands. This new influence was reflected in the creation, in 1964, of the United Nations Conference on Trade and Development and the Group of 77 'non-aligned' developing countries ('G-77') (the latter growing to 130 member states by the early twenty-first century).

Developing countries' demands for international economic justice were also manifested in the so-called New International Economic Order, a package of demands made during the 1970s and 1980s to restructure the world economy so that it would redistribute global financial resources to the developing countries. The package included calls for removal of barriers to trade, stabilisation of markets for basic commodity exports, increased funding from the International Monetary Fund and the World Bank, greater participation by the developing countries in decision making in international financial institutions, technology transfers from the developed countries, economic sovereignty (whereby the developing countries would control their own resources) and increased aid from rich to poor countries, at minimum equalling 0.7 per cent of donor countries' economic product (a target set by the United Nations). These demands did not meet with much success. In most respects the developing countries reduced their efforts to transform the world economy, instead focusing on meeting their demands in specific issue areas.

Where international justice became especially prominent was in international *environmental* relations. The debate surrounding the United Nations Conference on the Human Environment, held in Stockholm in 1972, occurred in the midst of these nascent but mostly ineffective calls by developing countries for greater international justice. Many of the demands reflected in the New International Economic Order were, to varying degrees, codified in the 1982 Law of the Sea, in amendments to the 1987 Montreal Protocol on Substances that Deplete the Ozone Layer, and in a number of other international environmental agreements negotiated or signed at the 1992 United Nations Conference on Environment and Development – the 'Earth Summit' (see below).

THE DOCTRINE OF INTERNATIONAL ENVIRONMENTAL JUSTICE

The notion of environmental (or 'ecological') justice – ensuring that environmental benefits and burdens are shared and distributed fairly within a country or its sub-regions – originated in the United States as a way of identifying and protesting against the disproportionate environmental burdens borne by poor and minority communities located near landfills and polluting industries (Bullard 1990; Pellow and Brulle 2005).

Today, environmental justice is increasingly viewed as an important consideration in domestic environmental policy making in most countries (Schlosberg 2007). Recognition that it is the poor and the weak that bear most of the burdens of environmental pollution has been applied to relations *among* states as well.

JUSTICE IN INTERNATIONAL ENVIRONMENTAL AGREEMENTS

Justice has played an important role in international environmental agreements in recent decades, and indeed it has become a norm – even a doctrine of sorts – in guiding contemporary international negotiations in this context. Environmental justice outside state borders has been premised on the rights and duties of states. Underpinning the idea of international environmental justice is an interest-based logic: the consequences of international environmental pollution, such as stratospheric ozone depletion, ocean pollution and climate change, can be limited or prevented only if both economically developed and large developing countries reduce their polluting emissions. Unilateral efforts by the developed countries, while essential, will be overwhelmed as the large developing countries use more energy and produce more environmental pollutants. For example, in the case of climate change, if China burns its vast coal reserves and Brazil and Indonesia cut their expansive forests, greenhouse gas levels will increase far beyond the control of the developed countries. While wealthy countries must do much more to reduce their emissions of greenhouse gases, the developing countries must be persuaded that they should forgo the energy-intensive industrialisation enjoyed by the developed countries, and instead develop in a manner that does not rely as heavily on fossil fuels. Such persuasion will require substantial concessions on the part of the developed countries, including redistribution of funds and technology.

Consequently, international environmental deliberations have become an important forum for discussions of international justice, at least insofar as considerations of fairness are incorporated into agreements to protect the global environment. Provisions for justice in Earth Summit agreements and related conventions, including the climate change convention, are found in calls for new and additional funds and technology transfers on preferential terms to help developing countries develop in a sustainable fashion, as well as changes to voting arrangements in international environmental funding institutions (for example, the Global Environment Facility) to give developing countries more authority in deciding how to allocate development assistance (Harris 1997b). Other conferences have tackled considerations of international justice (for example, the 1995 United Nations World Summit for Social Development). But it is in the environmental issue area where international justice has become prominent, in large measure because so many countries may suffer the consequences of global environmental change, and because the developed countries – which are best equipped to take effective action – are recognising the degree to which poverty and old-fashioned industrialisation in the poor countries can contribute to adverse environmental changes that can affect the developed countries' environmental security and economic vitality (Harris 1999b, 2001b).

It is difficult to think about environmentally sustainable development, from either an international perspective or a national one, without at least implicitly thinking about justice.[1] Unfair distribution of social, economic, political and environmental resources is often synonymous with unsustainable development. The poor are concerned about fulfilling their basic needs and, once that is accomplished, raising their living standards. They are unlikely to be immediately concerned with environmental changes whose adverse effects will be experienced or suffered in the relatively distant future, especially when those problems are largely caused by, and especially concern, the wealthy states and peoples of the world that the poor often blame for their suffering. The people of the developing countries (rightly) believe that it is unfair for the citizens of the developed countries to ask the poor to forgo development so that the developed countries can continue to consume as they have so far. Only if the poor are treated fairly by the rich will they genuinely join in efforts to protect the global environment. Thus, environmental change creates a situation in which interpretations of justice at the international level have greater salience than they might have without that environmental change (Harris 1996).

The next chapter describes how international justice has obtained in the context of climate change per se. In the following sub-sections I briefly describe how international justice has been manifested in some international environmental negotiations and agreements. This evolution of international environmental justice has served as the foundation for incorporating justice considerations into the climate change regime.

THE STOCKHOLM CONFERENCE

The 1972 United Nations Conference on the Human Environment held in Stockholm was the world's first nearly universal meeting among states to address environmental problems, with a majority of the world's governments participating. Before this conference, most international environmental meetings focused on scientific issues, but in Stockholm economic differences between rich and poor countries were major elements of the discussions. As would happen twenty years later at the Rio Earth Summit (see below), in Stockholm developed-country diplomats emphasised humankind's adverse impact on the environment, while developing-country participants focused on economic and social development. Developing countries generally laid the blame for much of the poverty and pollution in the developing world on practices by wealthy states that exploited poor countries. They feared that agreements coming out of the conference might have adverse effects on their own development. They worried that stricter environmental standards in the developed countries would raise the price of manufactured products, exacerbating already unfavourable terms of trade. They also worried that scarce development funds would be diverted away from economic development to environment-protection projects. These differences were bridged, at least in conference dialogue, by the notion that protection of the environment was an integral component of effective socioeconomic development (Caldwell 1990: 56).

During the conference the developing countries demanded sovereignty over their biological resources, technology transfers from rich to poor countries, and access to additional financial resources. The most divisive topic of deliberation was the first demand: that developing countries be allowed to share in the economic benefits of biotechnology, such as pharmaceuticals, derived from biological resources taken from their territories. The developing countries began to connect their demands for technology transfer to access to biological diversity by developed countries. Diplomats agreed at the conference that all states should have sovereignty over their biological resources. The developed countries were also persuaded to include other issues of concern to poor countries in conference declarations. These included provisions for improving access to food and clean drinking water in developing countries, as well as the broader development concerns of poor countries.

A number of principles related to international justice appeared in the Stockholm Declaration (United Nations 1972). Paragraph 4 of the declaration stated that environmental problems in the developing countries are caused primarily by underdevelopment, that millions of people there live 'far below minimum levels required for a decent human existence, deprived of adequate food and clothing, shelter and education, health and sanitation', and therefore 'the industrialized countries should make efforts to reduce the gap between themselves and the developing countries'. Principle 9 stated that 'environmental deficiencies generated by the conditions of underdevelopment and natural disasters pose grave problems and can best be remedied by accelerated development through the transfer of substantial quantities of financial and technological assistance'. Principle 12 called on the developed countries to take into account the particular requirements of the developing countries and 'any costs which may emanate from their incorporating environmental safeguards into their development planning and the need for making available to them, upon their request, additional international technical and financial assistance for this purpose'. Principle 21 declared that countries have the 'sovereign right' to exploit their own resources as they choose, and the 'responsibility to ensure that activities within their jurisdiction or control do not cause damage to the environment' of other countries.

These statements were early steps towards incorporation of international justice into environmental agreements among states. Without developing-country efforts at the Stockholm conference, the meeting would have focused on the environmental agenda of developed states, including pollution, population growth, resource conservation, limits to growth and the like, rather than relationships between environment and economic development (M. Miller 1995b). Overall, the Stockholm conference demonstrated a greater awareness of international justice as it relates to the environment. However, beyond this, there was little that emerged from Stockholm by way of significant *implementation* of international environmental justice.

THE LAW OF THE SEA

The Third United Nations Conference on the Law of the Sea was initiated in 1973 by the United States and the Soviet Union to protect freedom of navigation at a time when developing coastal states were claiming sovereignty and economic rights to larger areas of adjacent waters. The resulting Law of the Sea treaty (United Nations 1983), signed by nearly 160 countries in 1982 and entering into force in 1994, included a number of provisions for international justice. The treaty was negotiated during the heyday of the New International Economic Order, which made developed countries wary of developing-country proposals for collective ownership of the deep seabed, global taxes, and technological and financial transfers. Nevertheless, the treaty provided for sharing of mineral resources on the seabed outside the maritime boundaries of coastal states. The deep ocean floor beyond the exclusive economic zones of coastal states was declared the 'common heritage of mankind'. These areas were to be managed by the International Seabed Authority and an international corporation called the Enterprise. Developing countries hoped the seabed authority's exclusive rights to the deep seabed would produce new financial resources to promote their development.

The developed maritime powers essentially traded transit passage through international straits and freedom of navigation on the high seas for the developing countries' demands for extended exclusive economic zones and international control of deep seabed mining (Larson 1994). Demands by developing countries that the deep-ocean seabed be declared the common heritage of mankind had to be taken seriously if maritime powers were to ensure predictable access to coastal waters and straits of passage that could be restricted if developing littoral states declared those areas sovereign territorial waters. This was a situation in which the developing countries – at least coastal states among them – had newfound bargaining leverage that persuaded developed countries to take demands for international justice more seriously. The developed countries were faced with a clash between their national objectives and the unique characteristics of ocean 'commons'. Like other global commons – such as the ozone layer, the atmosphere and outer space – the world's oceans can benefit all users if managed cooperatively. But if a few countries choose not to cooperate because they believe that doing so would promote their own short-term objectives, many states – in this case the maritime developed countries – might suffer. The developed states could not achieve their objectives without the participation of developing countries.

According to some observers, the Law of the Sea is the closest the developing states have come to constructing their 'ideal' regime, in part because the treaty 'gave developing countries ready access to decision-making forums and invested them with more influence and power than they could ever have claimed on the basis of their national power capabilities' (Krasner 1985: 230). That being said, the Law of the Sea has produced no significant economic benefits from the deep seabed for poor countries, in large part because exploitation of resources there is rarely economically fruitful and the technology to take advantage of deep ocean resources remains in the hands of a few corporations in developed countries. Nevertheless, the provisions in the treaty for meeting some of the developing countries' demands for international justice reflected, and perhaps bolstered, the gradual change in attitude towards international environmental justice among states.

THE MONTREAL PROTOCOL

Multilateral efforts to prevent depletion of the Earth's stratospheric ozone layer began with the 1985 Vienna Convention for the Protection of the Ozone Layer and the follow-on 1987 Montreal Protocol on Substances that Deplete the Ozone Layer (Benedick 1998). The Vienna Convention focused on cooperation in the gathering of information.

Initial negotiations on ozone depletion were relatively easy because it was believed that participation of most countries was not essential for an agreement to be effective. It was thought that the Vienna Convention could be successful with participation of only those countries that were major producers of ozone-destroying chemicals. Thus the developing countries were not significant actors in shaping the convention. However, developing countries took a greater interest in negotiations for the Montreal Protocol, an agreement to limit the production of chlorofluorocarbons and other ozone-destroying chemicals. They feared that the agreement might limit their access to those chemicals, which were used in increasing quantities for refrigeration and other important purposes. In the event of restrictions on ozone-destroying chemicals, the developing countries wanted concessional access to substitute chemicals or financial assistance to help in purchasing them. China and India were especially vocal in their insistence that they should not have to suffer from efforts to fix a problem caused by the industrialised states (M. Miller 1995b: 73). At least in the early stages of negotiations for the Montreal Protocol, most developed countries were highly resistant to such demands.

Nevertheless, the protocol did incorporate some provisions to persuade the developing countries to join. For example, they were permitted to expand their use of chlorofluorocarbons during a ten-year transitional period. They were also entitled to transfers of new technology that would help them make the transition to new alternatives to ozone-destroying chemicals. However, the protocol did not include funding to help defray the costs incurred by the poorer countries making the transition to these chemicals. The result was that the protocol did not have developing-country support, with most of them refusing to sign it. Developing countries made their participation contingent on creation of a fund that would help finance their transition to the new, less harmful – and more costly – chemicals. By the time of the 1989 first meeting of the parties to the Montreal Protocol, developed countries agreed to only modest measures to help developing countries acquire information, research and training, and to aid them in their efforts to garner financing for technology transfers and the retooling necessary to fulfil obligations of the protocol.

However, in the light of increasing scientific knowledge about the dangers of ozone depletion, along with growing public concern in developed countries, by the early 1990s there was a new emphasis on completely phasing out ozone-destroying chemicals rather than just limiting their production. Doing this would require participation of most developing countries. Consequently, at the 1990 second meeting of the parties in London, developed countries agreed to substantial new efforts to bring the developing countries, especially the largest ones with the most potential to derail efforts to phase out ozone-destroying chemicals, into the protocol. The London amendments to the protocol contained several provisions for international justice that were absent from the 1987 agreement. For example, they 'acknowledge that special provision is required to meet the needs of developing countries, including provision of additional financial resources and access to relevant technologies' (United Nations Environment Programme 1990: preamble, para. 7). They established that developing countries have special needs and that those countries' compliance with the treaty will depend on funding and technology transfers from the more affluent parties (arts. 5(1) and 5(5)). The amendments called on parties to establish a multilateral fund 'for the purposes of providing financial and technical cooperation, including transfer of technologies' to developing countries to help them comply with the treaty. Contributions to the financial mechanism were 'additional to other financial transfers' and were to 'meet all agreed incremental costs' incurred by developing countries (art. 10(1)). Funds were to be provided to poor countries on a grant or concessional basis (art. 10(3)a). Under the amendments, the developed countries agreed to 'expeditiously' transfer applicable technologies to the developing countries 'under fair and most favourable conditions' (art. 10A(B)). Towards this end, the ozone Multilateral Fund was established during the early 1990s. It was the first fund dedicated to an international environmental agreement.

THE EARTH SUMMIT

Much like the Stockholm conference two decades earlier, the United Nations Conference on Environment and Development, which met as the Earth Summit in Rio de Janeiro in 1992, was initiated by developed countries concerned about the environmental consequences of industrialisation, and early preparations for the conference focused on those countries' objectives (M. Miller 1995a: 249). However, the differences between rich and poor states witnessed at Stockholm were also apparent: the former wanted to focus on environmental problems, whereas the latter wanted to emphasise economic development. As the date for the summit neared, the sentiments of the developing countries – that environmental protection was impossible while significant international injustices prevailed – became much more salient (M. Miller 1995b: 9).

The theme of the summit – environmentally sustainable development – emerged from the World Commission on Environment and Development, or Brundtland Commission. In the Brundtland Report

(WCED 1987), the commission highlighted the links between poverty, development and environment. It defined sustainable development as economic development that meets the needs of present generations without impeding future generations from meeting their own needs. The Brundtland Commission was explicit in stating that the concept of sustainable development must encompass efforts to meet the essential needs of the world's poor, 'to which overriding priority should be given' (WCED 1987: 43). The commission argued that it is 'futile to attempt to deal with environmental problems without a broader perspective that encompasses the factors underlying world poverty and international inequality' (WCED 1987: 3). The concept of sustainable development, because it brought together notions of ecology, economic development and poverty, set the stage for more serious incorporation of justice into international environmental diplomacy.

The General Assembly established the United Nations Conference on Environment and Development in a resolution adopted at the end of 1989 (United Nations 1989). Echoing the Brundtland Report, the resolution called for an international conference that would address environmental protection *and* economic and social development. The resolution was permeated – like the resulting agreements made at Rio – with provisions for international social justice. For example, it declared that poverty and environmental degradation are closely interrelated, that developing countries have special needs, and that the 'promotion of economic growth in developing countries is essential to address problems of environmental degradation' (para. 5). It also declared that developed countries were most responsible for destruction of the environment: 'the largest part of the current emissions of pollutants into the environment, including toxic and hazardous wastes, originates in developed countries, and therefore . . . those countries have the main responsibility for combating such pollution' (para. 9). The resolution called for 'favourable access to, and transfer of, environmentally sound technologies, in particular to developing countries, including on concessional and preferential terms [and] assured access of developing countries to environmentally sound technologies' (para. 15(m)), and the resolution declares in several places that 'new and additional financial resources will have to be channelled to developing countries in order to ensure their full participation' (preamble). The resolution also calls on governments to create a special international fund 'with a view to ensuring, on a favourable basis, the most effective and expeditious transfer of environmentally sound technologies to developing countries' (para. 15(1)).

There were 178 states participating in the Earth Summit, making it at the time the largest international conference ever held. Results of the summit included the Rio Declaration on Environment and Development; Agenda 21, a lengthy policy statement; a convention on biological diversity and the climate change convention; a statement on forest principles; establishment of a new United Nations Commission on Sustainable Development; and expansion of the Global Environment Facility.[2] To varying degrees, all of these products of the Rio conference incorporated provisions that fit many of the conceptions of justice described in the previous chapter, particularly when compared to most international agreements in other issue areas. These provisions included calls for concessionary or preferential technology transfers, soft loans, and new and additional funding for developing countries, in line with the General Assembly's earlier resolution.

The Rio Declaration on Environment and Development (United Nations 1993) is indicative of the mood at the Earth Summit. It contains several provisions for international justice. In its first principle, it acknowledges (in surprising cosmopolitan fashion) that individuals are at the centre of concern for sustainable development, and in principle 3 it states that 'the right to development must be fulfilled so as to equitably meet developmental and environmental needs of present and future generations'. Principle 5 declares that 'all States and all people shall cooperate in the essential task of eradicating poverty as an indispensable requirement for sustainable development, in order to decrease the disparities in standards of living and better meet the needs of the majority of the people of the world'. Governments represented at the Rio conference also declared that the 'special situation and needs of developing countries, particularly the least developed and those most environmentally vulnerable, shall be given special priority' (principle 6) and that states have 'common but differentiated responsibilities', meaning that developed countries have a greater responsibility to take steps to protect the global environment and to help poorer countries do likewise (principle 7). Thus the Rio Declaration says that 'developed countries acknowledge the responsibility that they bear in the international pursuit of sustainable development in view of the pressures their societies place on the global environment and of the technologies and financial resources they command' (principle 7), a provision that appears, like the others, in the climate change convention (see Chapter 4).

THE BIODIVERSITY CONVENTION

The Convention on Biological Diversity (United Nations 1992) provides a good case study of how calls for justice among states – that is, among rich and poor states – have been codified in international environmental agreements. The convention was signed at the Earth Summit by 153 countries and the European Community, and it entered into force in 1993.[3] In convention negotiations, developing countries demanded full sovereignty over their genetic resources, much as they had done at Stockholm. They wanted concessional transfers of technologies developed from their genetic resources. In addition, they insisted on royalties for access to their biological resources and creation of a fund to help them meet the provisions of the convention. Developing countries viewed the convention as part of a larger agenda to use the Earth Summit negotiations as a vehicle for restructuring international economic relations, thus enabling them to acquire resources, technologies and market access that would bolster their economic development (Boyle 1994: 113–14). For the developed states, the main concern was protecting the pharmaceutical and biotechnology industries' intellectual property and access to plants, animals, microbes and other biological resources in developing countries (Chatterjee and Finger 1994: 43).

The objectives of the Biodiversity Convention, resulting from compromises intended to satisfy both developed and developing countries, are the 'conservation of biological diversity, the sustainable use of its components, and the fair and equitable distribution of the benefits arising out of the utilisation of genetic resources, including appropriate access to genetic resources and by appropriate transfer of relevant technologies, taking into account all rights over those resources and to technologies, and by appropriate funding' (United Nations 1992: art. 1). The convention's preamble acknowledges that 'the provision of new and additional financial resources and appropriate access to relevant technologies can be expected to make a substantial difference in the world's ability to address the loss of biological diversity', and that 'special provision is required to meet the needs of developing countries, including the provision of new and additional financial resources and appropriate access to relevant technologies'. The convention obligates parties to take steps to share 'in a fair and equitable way the results of research and development and the benefits arising from the commercial and other utilization of genetic resources with the Contracting Party providing such resources' (art. 15(7)). Similar provisions are made for technology transfer (art. 16(3)). Article 16 (para. 2) states that technology access and transfers to developing countries 'shall be provided and/ or facilitated under fair and most favourable terms, including on concessional and preferential terms where mutually agreed'. Parties to the treaty are required to 'take all practicable measures to promote and advance priority access on a fair and equitable basis by Contracting

Parties, especially developing countries, to the results and benefits arising from biotechnologies based upon genetic resources provided by those Contracting Parties' (art. 19(2)).

Article 20 outlines provisions for financial resources. It states that the 'developed country parties shall provide new and additional resources to enable developing country Parties to meet the agreed full incremental costs to them of implementing measures which fulfil the [convention's] obligations' (art. 20(2)). Article 20 also declares that the extent to which developing countries will effectively implement their commitments according to the convention will 'depend on the effective implementation by developed country Parties of their commitments related to financial resources and transfer of technology and will take fully into account the fact that economic and social development and eradication of poverty are the first and overriding priorities' of the developing countries (art. 20(4)). Parties to the treaty are to take 'full account of the specific needs and special situation of least developed countries in their actions with regard to funding and transfer of technology' (art. 20(5)).

It is apparent that international justice was an important, even central, theme of the Biodiversity Convention. It could be argued that, at the Earth Summit, developing countries were successful in gaining additional recognition of sovereign rights over biological resources within their territories. The Biodiversity Convention's requirements that access to genetic resources must be subject to the prior informed consent of the country where the collection occurs, and that access to such resources must be on mutually agreed terms, were new international law (Burhenne 1992: 325). This is the outcome one would expect from negotiations between states operating in an international system where sovereignty and non-intervention are the fundamental and most cherished components of the system, although not necessarily the result one might expect in a system where the most powerful states – the economically developed ones – traditionally get their way. In the case of the Biodiversity Convention, demands for international justice swung in the direction of the developing countries' priorities. According to Alan Boyle (1994: 116), a 'trade-off between conservation and economic equity is at the heart of the Convention and makes it unusual among environmental agreements'. However, the convention, and indeed other international environmental agreements, 'affirm and give precedence to the *physical* rights of states to their resources and the *authority* rights of states over how those resources can be used and exploited' (Elliott 2006: 353). Despite the convention, it is evident that, with very few exceptions, efforts to protect the Earth's biodiversity have failed. The pace of extinctions is *accelerating* as a consequence of habitat destruction, over exploitation, pollution and indeed climate change, with the rate of species loss now about one thousand times the natural rate (Speth 2008: 36–8). Thus, as Derek Heater (1996: 143) asks, 'the question must be raised whether nation-states, individually or collectively, can be trusted to pursue policies in the interests of the planet as a whole. At the root the problem has been caused by the pursuit of wealth.'

The Biodiversity Convention suggests that developing countries demonstrated some 'environmental power' at the Earth Summit, explaining to a great extent why justice found its way into international environmental negotiations and agreements there and subsequently. International *environmental* justice has its origins as an outgrowth of the general historical norms of international justice and demands by developing countries for justice in the context of environmental change and pollution, as well as an attempt to further their more general demands for a more equitable international economic order. We might conclude that the codification of international environmental justice has been a step forward for international justice. But it has been a very small step. For the most part, the developed countries have failed to implement the doctrine of international environmental justice that has been codified in international environmental agreements. This is due in large part to fundamental flaws in the doctrine itself. Because it is premised on states, which are characteristically self-interested in their behaviours and seldom willing or able to put morality first, and even routinely fail to see that doing what is right is in their best long-term interests, the doctrine's implementation is bound to be flawed. States have addressed transboundary environmental and natural resource problems in the same way that they have addressed other issues: by devising and agreeing to treaties and instruments that promote their individual and collective interests, with the Westphalian norms prevailing. Related questions of justice have normally involved notions of rights and obligations of *states*. Wolfgang Sachs (2002: 14) characterises the sustainable development project manifested in the Rio Earth Summit as a 'development-as-growth philosophy' that has been exploited out of political expediency by both developed and developing states. This focus on economic growth rather than environmental protection meant that 'the elites in the South and the North could reconcile themselves with the outcome of Rio. Indeed, it was an unholy alliance between Southern and Northern governments in favour of development-as-growth that has largely emasculated the spirit of Rio' (Sachs 2002: 14). The result was that the 'health of the planet further deteriorated and global inequality increased' (Sachs 2002: 15).

CONCLUSION

Many of the provisions of international environmental agreements reflect some of the interpretations of justice described in relation to climate change in Chapter 2. For example, states have recognised, at least in these treaties, that rich countries are most responsible for environmental problems and thus have some special duties to act first and to take other steps based on that responsibility. There is also clear intent in most international environmental agreements to help the least well-off countries. However, the doctrine of international environmental justice comes in different forms. We can identify what might be called – on one end of a spectrum – *selfish* doctrine. According to this interpretation, international environmental justice has come about and been codified in international environmental agreements simply as in instrument for states, especially powerful and wealthy ones, to promote their interests. They may have intended it to help foster international environmental cooperation in which they have an interest, but the doctrine has been implemented in a way that usually promotes the economic interests of developed states and multinational companies based there. On the other end of the spectrum one might identify what could be called *altruistic* doctrine. In this account, states have formulated international environmental justice and codified it in agreements among themselves in a genuine attempt to protect global commons and, importantly, to right past wrongs (especially historical pollution) and to aid the poor states of the world. There will be different interpretations of the doctrine at different points along the spectrum, or perhaps even more extreme interpretations, although one is hard pressed to find anyone to endorse an even more altruistic doctrine that aims to promote the interests of poor people regardless of where they live. What is important for our discussion is that the different forms of the doctrine are nearly *always about states*.

Two themes that are evident in the preceding description of justice in international environmental affairs are, first, that justice has been codified in agreements as a consequence of bargaining among states aiming to promote their own interests. The common interests they share in environmental issues frequently explain justice provisions of international environmental agreements: justice is a means to an end. However, because states' common interests also often conflict, justice is not implemented nearly as often as the agreements would lead one to believe states intend it to be. A second theme is related to the first: environmental protection is, for many states, itself also a means to an end. Justice depends on each state's particular perspective, usually defined in terms of economic growth at home. Consequently, the state-centric character of justice beyond borders ends up undermining both justice and environmental protection. As Patrick Hayden (2005: 132) reminds us, the Earth Summit agreements were attempts to enshrine the concept of sustainable development in international practice premised first on the maintenance of state sovereignty over national resources. He argues that, in the battle between state sovereignty and 'the moral responsibility enunciated in international environmental instruments that states not cause environmental damage which reaches beyond their own borders', sovereignty has triumphed (Hayden 2005: 132). He therefore questions 'whether a system which privileges state sovereignty over global environmental concerns can adequately address the problems facing the planet and its inhabitants as a whole, and concomitantly, whether such a system can lead to the actualization of justice across state boundaries' (Hayden 2005: 132).

Thus, one conclusion could be that the Earth Summit failed because 'the economic interests of the rich and poor nations could not be reconciled and subsumed under the greater good of the world as a whole' (Heater 1996: 144). Hayden (2005: 125) has argued that 'any conclusions concerning the idea of sustainable development remain subordinate to a framework which favours state and commercial interests'. He points to the second and third principles of the Rio Declaration, which declare that states have full sovereignty over natural resources within their borders and full rights to exploit them for economic gain (Hayden 2005: 125). As Robin Attfield (2005: 41) observes, communitarianism raises few objections to 'certain widespread obnoxious practices', such as the export to developing countries of hazardous wastes. He concludes from this that, 'if policies continue to be determined by communitarian rather than universalist principles, there is a danger that the environmental health of economically vulnerable countries of the Third World, and of coming generations in those countries, will continue to be endangered' (Attfield 2005: 41). In short, preoccupation with the Westphalian norms has undermined environmental protection *as well as* both international and global justice.

Given the imperatives of environmental protection generally and climate change in particular, perhaps it is time at least to consider more seriously the possibility that 'egoistical state sovereignty is now potentially lethal' (Heater 1996: 165). The codification of environmental justice in international relations is a positive step towards the actualisation of international justice. However, it has not been enough to address transnational and global environmental problems effectively. Indeed, by focusing all attention on the rights and obligations of *states*, international environmental justice may be acting as an *obstacle* to a truly effective agreement on climate change, as we will see in the next chapter. Perhaps the climate change regime has made progress in realising international environmental justice. But the effectiveness of the regime may depend on going further to realise *global* justice as well.

NOTES

1. This is not to assume that the justice and environmental agendas are always in common cause (see Dobson 1998).
2. The biodiversity and climate change conventions were signed at the Earth Summit, but they were not formally part of the same diplomatic process.
3. The United States was the only country at the Earth Summit not to sign the convention.

CHAPTER 4

INTERNATIONAL JUSTICE AND CLIMATE CHANGE

Like other international environmental agreements, those that comprise the climate change regime have been premised on Westphalian norms of state sovereignty and states' rights. The climate change regime has been guided in large measure by the doctrine of *international* environmental justice. The norms, discourse and thinking associated with this state-centric doctrine have taken the politics and diplomacy of climate change in a direction that has been characterised by diplomatic delay, minimal action – especially relative to the scale of the problem – and mutual blame between rich and poor countries, resulting in a 'you-go-first' mentality that has prevailed even as global greenhouse gas emissions have exploded. The problems that have prevented more effective action in other environmental areas have also been manifested in most domestic and international responses to climate change, all of which have been preoccupied with protecting perceived national interests. This focus on the rights and interests of states has been written into the climate change agreements, including the 1992 Framework Convention on Climate Change, the 1997 Kyoto Protocol and subsequent agreements and diplomatic negotiations on implementing the protocol and devising its successor. Although some major industrialised countries, notably in Europe, have started to restrict and even reduce their emissions of greenhouse gases, these responses pale in comparison to the major cuts demanded by scientists. Just as profoundly, many large developing countries are experiencing huge emissions increases as their economies grow, often bolstered by increased exports, and as millions of their citizens finally escape poverty and join the global middle class. These growing emissions will dwarf planned cuts by developed states. In short, most of the world's expanding wealthy classes continue to consume and pollute aggressively, usually without any legal restrictions, regardless of the growing impact on the planet.

Building on previous chapters' descriptions of the problem of climate change (Chapter 1) and the incorporation of justice into international environmental agreements (Chapter 3), this chapter briefly describes the formation of the climate change regime, its provisions for international justice, and its effects so far. As we will see, the climate change regime may be at or near the zenith of justice in international environmental affairs, again demonstrating that most states recognise the need for justice in this issue area. However, while the regime may be considered an important step for international justice, and again shows the power of developing countries in the environmental context, it has not achieved the regime's aim of preventing dangerous impacts on the Earth's climate system. Indeed, the causes and consequences of climate change are growing much worse.

THE CLIMATE CHANGE REGIME

One of the first major international events dealing with climate change was the 1979 First World Climate Conference, a gathering of scientists interested in climate change and its relationship with human activities. From that conference a programme of scientific research was established, leading to the creation of the Intergovernmental Panel on Climate Change in 1988. The intergovernmental panel's first assessment report and the Second World Climate Conference in 1990 added stimulus to initial concerns about climate change among governments. Consequently, in December 1990 the United Nations General Assembly established the Intergovernmental Negotiating Committee for a Framework Convention on Climate Change, which was the basis for subsequent international agreements on the issue. From then until the Earth Summit, representatives of over 150 countries negotiated the Framework Convention on Climate Change (the climate change convention). The convention was signed by 153 countries plus the European Community. It came into effect in 1994. The stated aim of the convention is the 'stabilization of greenhouse gas concentrations in the atmosphere at a level that would prevent dangerous anthropogenic interference with the climate system' (UNFCCC 1992: art. 2). It called on the world's most economically developed countries *voluntarily* to reduce their emissions of greenhouse gases to 1990 levels by 2000. That objective was not achieved. Most developed countries were ready to accept the reduction target as a binding commitment, but only if the United States agreed to do so. It did not.

In 1995 parties to the climate change convention began a series of periodic conferences. At the first conference of the parties, held in Berlin in 1995, developed countries acknowledged that they had a greater share of the responsibility for causing climate change and they agreed to act first to address the problem. In line with demands from developing countries, the meeting culminated in the so-called Berlin Mandate, in which the industrialised countries agreed to take on greater commitments to reduce their greenhouse gas emissions and to assist poor countries with sustainable development.[1] This first conference affirmed the notion of 'common but differentiated responsibility', meaning that all states have a common responsibility to address climate change, with the proviso that developed countries have greater ('differentiated') obligation to do so (see below).

At the second conference of the parties, held in Geneva in 1996, governments called for a legally binding protocol with specific targets and timetables for reductions of greenhouse gas emissions by developed states. The resulting Geneva Declaration served as the negotiating basis for a protocol to the climate change convention, which was agreed in December 1997 at the third conference of the parties in Kyoto. The Kyoto Protocol requires most developed-country parties to reduce their collective greenhouse gas emissions to 5.2 per cent below 1990 levels by 2012. However, not all of them agreed to be bound by the protocol, with the United States being the most important hold-out. The Kyoto conference proved to be especially contentious, not least because the United States seemed to be reneging on the Berlin Mandate when it called for the 'meaningful participation' of developing countries. Nevertheless, the protocol established specific emissions goals for developed countries without requiring significant commitments from developing countries. The protocol also endorsed emissions-trading programmes that would allow developed countries to buy and sell emissions credits among themselves. Other so-called flexible mechanisms included in the protocol were 'joint implementation', whereby developed countries could earn emissions credits when investing in one another's emissions-reduction projects, and the Clean Development Mechanism, which allows developed-country entities to pay for, and receive emissions credits for, emissions-reduction projects in developing countries.

Negotiations on how to implement the Kyoto Protocol took place over a number of years. Some of the means by which the protocol's 5-per-cent goal would be reached were codified at the 1998 fourth conference of the parties in Buenos Aires. At the fifth conference in Bonn in 1999, parties agreed to a timetable for completing outstanding details of the protocol. The sixth conference of the parties began in November 2000 in The Hague, but talks broke down because of disagreements among delegates, particularly on the question of carbon sinks. The Kyoto Protocol's ratification was put in doubt with the advent in early 2001 of President George W. Bush in the United States. He withdrew all United States support for international negotiations to address climate change seriously. The sixth conference of the parties resumed in Bonn during July 2001. The resulting Bonn Agreement clarified plans for emissions trading, carbon sinks, compliance mechanisms and aid to developing countries. At the seventh conference of the parties in Marrakech later in 2001, diplomats negotiated the Marrakech Accords, a complicated mix of proposals for implementing the Kyoto Protocol that were largely designed to garner ratification from enough states to allow the protocol to enter into force. Parties agreed to increase funding for the climate change convention's financial mechanism, the Global Environment Facility, as well as to establish three new funds that were intended to provide additional aid to poor countries: the Least Developed Countries Fund, the Special Climate Change Fund and the Adaptation Fund. Importantly, the parties to the accord agreed to the need to even out per-capita differences among developed and developing countries, in effect accepting the notion that all citizens of the world have equal rights to use the atmosphere (Sachs and Santarius 2007: 188).

At the October 2002 eighth conference of the parties in New Delhi, a tacit agreement between the United States, a few other developed countries and several large developing countries, notably China and India, emerged that shifted much of the focus away from mitigating climate change and towards adaptation. It was at this conference of the parties, and the one that followed, that diplomats discussed ways to implement the Marrakech Accords and to prepare for ratification of the Kyoto Protocol. The tenth conference of the parties, held in Buenos Aires in December 2004, was dubbed the 'Adaptation COP' because discussion there also focused more on adaptation to climate change than the more common discussions, before 2002, of lowering greenhouse gas emissions. The conference resulted in pledges of more assistance to aid poor countries most affected by climate change, but there were no firm commitments to ease access to major new funding for adaptation. It was also in 2004 that Russia ratified the Kyoto Protocol, allowing the agreement to enter into force in February 2005.

One visible aspect of the climate change negotiations has been the acrimony between the developed countries – particularly the United States – and the developing world. This was revealed during the late-2005 eleventh conference of the parties to the climate change convention and the first 'meeting of the parties' to the Kyoto Protocol, which were held simultaneously in Montreal. The meeting formalised rules for implementing the Kyoto Protocol (for example, rules for emissions trading, joint implementation, crediting of emissions sinks, penalties for non-compliance), strengthened the Clean Development Mechanism, began negotiations for further commitments by developed-country parties to the protocol beyond 2012 (when the protocol's commitments would expire) and set out guidelines for the Adaptation Fund. Several developing countries, while still opposed to binding obligations to limit their greenhouse gas emissions, showed interest in undertaking voluntary measures, in keeping with the principle of common but differentiated responsibility.

In his opening address to the November 2006 twelfth conference of the parties in Nairobi, United Nations Secretary-General Kofi Annan (2006) characterised the negotiations up to that point as displaying a 'frightening lack of leadership' from governments. By the next conference of the parties, held in Bali in 2007, discussions were pushed by the Intergovernmental Panel on Climate Change's fourth assessment report, which removed any remaining doubt about the seriousness of the problem. The meeting was important in its widespread opposition to efforts by American diplomats to thwart negotiations on a new, post-2012 agreement that would require developed states to take on new obligations to limit their greenhouse gas emissions and to aid developing countries with sustainable development. In the end, developing-country governments agreed that they would consider taking unspecified future actions to mitigate their emissions, which was a substantial shift from their long-standing policy of refusing to accept any commitments. The quid pro quo was a streamlining of the Adaptation Fund and resourcing it with a new levy of 2 per cent on Clean Development Mechanism projects. Developed countries also agreed to new emissions targets and timetables. However, as with the developing states' agreement, nothing was specified and there were no interim targets to test the willingness of governments to implement their commitments. Diplomats instead adopted the so-called Bali Roadmap, intended to guide discussions leading to a new, comprehensive agreement, under both the climate change convention and the Kyoto Protocol, to be negotiated in time for a conference of parties in Copenhagen at the end of 2009 (Pew Centre 2007).

This evolution of the climate change regime reveals the important role of international justice. As with other international environmental agreements described in the previous chapter, action on the justice concerns of developing countries was required to garner their participation in the climate change regime. The developed states came to recognise that the problem of climate change could not be fully addressed without the support of developing countries, whose involvement was overshadowed by long-standing concerns about international justice. Discussions of who was responsible for climate change emphasised 'the unique responsibility of the developed world and the sovereign countries it comprises' (Page 2008: 557). A focus on *national* responsibility served as a foundation for the climate change convention, the Kyoto Protocol, and the regime's broader guiding principle of justice: common but differentiated responsibility.

COMMON BUT DIFFERENTIATED RESPONSIBILITY

As a nascent principle of international environmental law, common but differentiated responsibility evolved from the notion of 'common heritage of mankind' that gained stature in the Law of the Sea, and from the international designation of certain areas (Antarctica and the deep seabed) and resources (for example, whales) as 'common interests' of humankind (Stone 2004). Bearing in mind that humans depend on a healthy climate for their survival, the General Assembly went further by recognising the Earth's climate as a 'common concern' of humankind, implying not only the need for international cooperation to protect human interests but also indicating a 'certain higher status inasmuch as it emphasizes the potential dangers underlying the problem of global warming [and implying] that international governance regarding those "concerns" is not only necessary or desired but rather essential for the survival of humankind' (Biermann 1996: 431). Bearing in mind that the climate is of such crucial common concern to all people, it follows that there is a responsibility to protect it. This begs the question of who is responsible. The answer has been a function of each state's historical contribution to the problem, its level of economic development and its ability to take action. This was suggested by the 1972 Stockholm Declaration, which stated that it is essential to consider 'the extent of the applicability of standards which are valid for the most advanced countries but which may be inappropriate and of unwarranted social cost for developing countries' (United Nations 1972: principle 23).

The principle of common but differentiated responsibility was described succinctly in the 1992 Rio Declaration:

> States shall cooperate in a spirit of global partnership to conserve, protect and restore the health and integrity of the Earth's ecosystem. In view of the different contributions to global environmental degradation, States have common but differentiated responsibilities. The developed countries acknowledge the responsibility that they bear in the international pursuit of sustainable development in view of the pressures their societies place on the global environment and of the technologies and financial resources they command. (United Nations 1993: principle 7)

According to this principle, while all states are responsible for global environmental problems, some are more responsible than others. The principle was also implicit in the Montreal Protocol, the Second World Climate Conference and other international deliberations (see Sands 1994).

The climate change convention followed these developments, declaring:

> The Parties should protect the climate system for the benefit of present and future generations of humankind, on the basis of equity and in accordance with their common but differentiated responsibilities and respective capabilities. Accordingly, the developed country Parties should take the lead in combating climate change and the adverse effects thereof. (UNFCCC 1992: art. 3(1))

The preamble to the convention notes that most current and historical emissions of greenhouse gases have originated in the developed countries and that 'per capita emissions in developing countries are still relatively low and that the share of global emissions originating in developing countries will grow to meet their social and developmental needs', and that actions to address climate change should first consider the 'legitimate priority needs of developing countries for the achievement of sustained economic growth and the eradication of poverty'. In short, states acknowledged that, while all of them should be part of efforts to limit emissions of greenhouse gases, the developed states would take the lead and they would help the world's poor countries address both the causes and the consequences of climate change.

The convention has other provisions for international justice. In addition to requiring developed countries to reduce their greenhouse gas pollution, it calls on them to give 'new and additional' resources (in addition to existing aid), in an 'adequate' and 'predictable' fashion, to assist developing countries in complying with their obligations under the agreement (UNFCCC 1992: art 4(3)). Developed countries are to take steps to 'promote, facilitate and finance, as appropriate, the transfer of, or access to, environmentally sound technologies and know-how', with the proviso that the developing countries' effective implementation of the treaty 'will depend on the effective implementation by developed country Parties of their commitments . . . related to financial resources and transfer of technology and will take fully into account that economic and social development and poverty eradication are the first and overriding priorities' of developing countries (UNFCCC 1992: art 4(5)).

Developing countries joined the convention only after it was agreed that their development prospects would not suffer in the process. Such an agreement included an implicit understanding that some sort of international fund would be established to compensate them for the costs of participation (Mott 1993: 302). Thus the convention described a financial mechanism that would provide funding 'on a grant or concessional basis, including for the transfer of technology' to help poorer countries fulfil treaty commitments, with the mechanism having 'an equitable and balanced representation' of parties to the convention 'within a transparent system of governance' (UNFCCC 1992: art. 11, paras 1–2). Details of the financial mechanism were explicitly put off to a later date. The developed countries wanted to control funds, in part to ensure their effective use. The developing countries wanted to participate in decision making regarding the dissemination of funds. The Global Environment Facility, jointly administered by the United Nations Development Programme, the United Nations Environment Programme and the World Bank, was designated as the interim financial mechanism, with the understanding that it would be 'appropriately restructured and its membership made universal' (UNFCCC 1992: art. 21(3)).

The assumption of the climate change convention was that it would be unfair to expect developing countries, especially the poorest among them, to limit their economic development when the wealthy countries of the world are most responsible for present concentrations of atmospheric greenhouse gases and the consequences of that pollution. Thus the convention is meant to achieve the objective of reducing greenhouse gas pollution to manageable levels in ways that are both effective and fair.

The principle of common but differentiated responsibility was reaffirmed in the Berlin Mandate, which was agreed at the first conference of the parties in 1995 (United Nations 1995). The mandate reminds governments that they are required to consider the special needs of the developing countries and:

> The fact that the largest share of historical and current global emissions of greenhouse gases has originated in developed countries, that the per capita emissions in developing countries are still relatively low and that the share of global emissions originating in developing countries will grow to meet their social and development needs. (United Nations 1995: art. I((1)(d))

> The fact that the global nature of climate change calls for the widest possible cooperation by all countries and their participation in an effective and appropriate international response, in accordance with their common but differentiated responsibilities and respective capabilities and their social and economic conditions. (United Nations 1995: art. I(1)(e))

Negotiations for the Kyoto Protocol between 1995 and 1997 were premised on the same principle. In keeping with this premise, developing countries vetoed any language in the protocol that would call on them to make even voluntary commitments to limit their emissions of green-house gases (Earth Negotiations Bulletin 1997). Accordingly, the protocol requires only *developed* countries to reduce emissions. In not requiring developing countries to limit, let alone reduce, their green-house gas emissions, the protocol conforms to the climate change convention's provisions for common but differentiated responsibility, specifically affirming the Berlin Mandate. Article 10 of the protocol makes this explicit when it states that all parties must take into account 'their common but differentiated responsibilities and their specific national and regional development priorities, objectives and circumstances, without introducing any new commitments for Parties not included in Annex I' (the list of developed-country parties to the convention).[2]

Among the chief concerns about international justice in the context of climate change are those regarding fairly allocating emissions (that is, mitigation) and uneven impacts (for example, adaptation), with the former pointing to the disproportionately high emissions from developed states and the latter highlighting the disproportionate – even perverse – suffering among poor countries who have (at least until recently) contributed the least to the problem (Park 2005: 178–9).[3] If one accepts several of the notions of justice outlined in Chapter 2, the developed countries ought to do more than cut their greenhouse gas emissions; they also ought to compensate those who suffer from the resulting effects of climate change. For example, the low-lying island and coastal states will suffer the effects of rising seas. They want to be – and ethically ought to be – compensated for these adverse effects. However, this is one area where both the climate change convention and the stated intentions and actions of the developed countries fall far short of many principles of justice. The most the treaty says about this is article 4(4), which declares that developed countries 'shall also assist the developing country Parties that are particularly vulnerable to the adverse effects of climate change in meeting costs of adaptation to those adverse effects'. Put another way, 'compensation is only formulated [in the convention] as a vague principle without any concrete implementation scheme' (Paterson 1997: para. 5.4.3.3).

The Bonn and Marrakech accords established several new funding mechanisms to help poor countries adapt to climate change. But funding for emissions-*mitigation* projects in developing countries has been quite small, usually on the order of $1 billion per year (but going as high as $10 billion in 2006), with most of that going to large-scale hydroelectric projects (Roberts et al. 2008). Meanwhile, funding for *adaptation* in developing countries has been 'minuscule', adding up to only about $600 million from 2000 to 2006, compared with estimates of how much is needed reaching $80 billion per year (that is, actual funding is about 99 per cent less than what is needed) (Roberts et al. 2008: 6).

Madeleine Heyward (2007) describes some perspectives on international justice ('equity') that have prevailed in the international negotiations on climate change. She identifies three core principles that have formed the basis of arguments by the states involved: equality, responsibility and capacity. Justifications for equality fall into three categories (Heyward 2007: 520): egalitarianism (all people have rights to equal shares of the atmospheric commons), sovereignty (all states have equal rights to the atmosphere) and comparability (all states ought to contribute equally to addressing climate change). The principle of responsibility is premised on two arguments: the polluter-pays principle (each state's responsibility is linked to its contribution for causing climate change) and benefit (responsibility to combat climate change is linked to benefits a state enjoys from doing so). The principle of capacity derives from four possible arguments: the economic situation and resource availability (states with the economic and resource capabilities ought to address climate change), basic needs (developing states' basic needs should be the main concern), domestic constraints (domestic weaknesses matter when deciding who has the capacity to act) and opportunities (capacity should be based on availability to a state of cost-effective actions).

One clear theme arising from Heyward's typology of equity principles is that arguments for justice are many, and each state can invariably identify (as states have done) one of them that promotes its perceived national interests. Indeed, sovereignty, along with comparability, is clearly reflected in the Kyoto Protocol (Heyward 2007: 525). Another theme, which will become important in Part III, is that only one of these principles shares the cosmopolitan focus on human beings: egalitarian equality. This latter principle has been invoked quite vociferously by developing countries, which argue that emissions targets should be based on each country's population. Thus, even this cosmopolitan-like principle is expressed in statist terms. However, Heyward (2007: 521) believes that egalitarian principles are unlikely to be politically viable in the climate change regime because those states that would bear massive costs in implementing them – the developed countries with relatively small populations – would not agree to egalitarianism as the basis for action. The cosmopolitan corollary to international environmental justice (see Chapter 7) proposes a way around this political problem (of a cosmopolitan principle being undermined by selfish states) by distributing the costs among affluent *people* everywhere, and by redistributing some funds even to developed states, specifically to some poor people there, although most of the people benefiting from funding would be those living in poor countries.

As the poor and less affluent countries of the world develop economically, they are becoming major sources of greenhouse gases, in aggregate now surpassing the developed states (Global Carbon Project 2008). Consequently, in the long run the developed countries cannot by themselves solve this problem; comprehensive participation of the developed countries *and* the major developing countries is required. This will have to include emissions limitations among developing countries. However, the world's poor countries are unlikely substantially to reduce their emissions simply because the wealthy countries want them to do so. As P. R. Shukla (1999: 157) puts it, 'the basic concern of the developing countries is not whether or when to initiate the mitigation actions, but how the mitigation burden will be distributed among nations. This is a justice issue, concerned with an equitable distribution.' Consequently, for international climate change negotiations to be successful in coming years – for them to result in actual, robust and widespread overall cuts in greenhouse gases – the developed countries must *demonstrate* that they acknowledge their 'guilt' for the damaging greenhouse gases they have emitted for centuries, and that they are acting substantively to reduce their future emissions. In addition, in accordance with the climate change convention, the developed countries will have to do much more to assist developing countries in their efforts to deal with this problem.

According to Wolfgang Sachs and Tilman Santarius (2007: 147), adoption of the principle of common but differentiated responsibility 'is the first sign that the legal significance of past actions for the present has basically been recognised. For, even if it was not intended as such, what looks like a diplomatic concession may be read as a confession that over the last few centuries the North has accumulated ecological debts to the South, by consuming parts of the environment the South now lacks for its development.' Common but differentiated responsibility, while perhaps not yet considered a customary norm of international law, now serves as a guiding principle for creating, interpreting and applying environmental conventions (Matsui 2002). But, as Yoshiro Matsui (2002) shows, the principle is flawed in practice because, instead of encouraging necessary changes in consumption and production patterns, developed countries are obliged only to assist developing countries with the incremental costs associated with implementing the climate change agreements, rather than assisting them to attain environmentally (and atmospherically) sustainable economic development.

Chukwumerije Okereke (2008: 99) argues that 'it is the conceptions of justice that have a close fit with the dominant neoliberal economic system which actually underwrite the core policies' of the climate change regime. He believes, for example, that even the Clean Development Mechanism has been transformed into a profit-making venture for the benefit of developed states and corporations based in those states, resulting in most investments by the mechanism being made in places 'that can guarantee quick financial returns rather than in areas that are the most vulnerable to the threat of climate change' (Okereke 2008: 99). Indeed, it has been shown that in practice the Clean Development Mechanism can result in *developing* countries financing developed countries' mitigation strategies because developing country companies may finance the projects before selling the resulting emissions credits (Lutken and Michaelowa 2008). From this perspective, the climate change regime is merely an instrument for narrow developed-country interests to further their domination of the global economy. Climate change mitigation and adaptation end up being not the objective at all, at least not in implementation, but instead have been co-opted by economic elites. Okereke (2008: 121–2) acknowledges that the 'determination of developing countries' led to 'greater procedural and substantive justice' in the regime, but he believes that this has been 'effectively co-opted for neoliberal ends . . . which not only have the potential of allowing developed countries to increase actual emissions but also the opportunity to strategically increase their capital base while creating the impression that they are helping the developing countries'.

There may be much to support Okereke's criticisms of the climate change regime. The argument that the outcome has not been more environmentally sustainable, in the form of major cuts in greenhouse gases by the developed countries and implementation of the various justice (and other) provisions of the regime, is hard to discount. Although climate change diplomacy since the late 1980s has been successful in garnering substantial participation in the regime among developing states, and has been the vehicle for bringing economic justice into international relations possibly like no other issue, it has done far too little to address the causes and consequences of climate change – or of international injustices. One might even argue that the provisions for justice that have been written into the agreements mostly serve as means to garner political support for a weak regime. The greenhouse gas cuts that have been actualised are absolutely tiny compared to the scale of the problem, and even the larger cuts pledged by some developed states in recent years are looking increasingly inadequate, even if we assume that they will be fully implemented. Indeed, the problem itself is much worse than even the worst predictions of the Intergovernmental Panel on Climate Change (see Chapter 3), as the next section reveals.

THE TRAGEDY OF THE ATMOSPHERIC COMMONS

All the international efforts to address climate change have been grossly inadequate when viewed in relation to the severity of the problem. Past and present greenhouse gas emissions mean that serious climate change is inevitable. Despite provisions in the climate change agreements for international justice, designed in large part to garner participation by more states, they have failed to bring about substantial changes in behaviour relative to the scale of the problem. This is a typical tragedy of the commons, albeit on a truly monumental scale. Fundamental to this tragedy of the atmospheric commons is the problematic 'you-go-first' mindset among most states, which has resulted in only modest greenhouse gas cuts in Europe (as Europe has sought to demonstrate some leadership, with the hope that others will follow) and no cuts at all in most of the developed world, not to mention continuing increases in the developing countries and globally (because increases in developing countries far surpass cuts in developed countries, meaning there will be global increases well into the future). Even with full implementation, the Kyoto Protocol would result in reductions of well under 5 per cent of parties' emissions because the manner in which countries are allowed to meet their commitments (for example, emissions trading and land-use changes) often will not result in significant actual cuts.

Insofar as the climate change regime is about reducing emissions of greenhouse gases, it is about mitigating future global warming rather than realistically halting, let alone reversing, it. Even *stabilisation* of greenhouse gas concentrations in the atmosphere would require massive cuts, at minimum 80 per cent below current levels but probably very much more (Speth 2008: 29). Scientists argue that emissions of carbon dioxide must be ended *completely* just to stabilise atmospheric concentrations and prevent chaos in the global climate system (see Mathews and Caldeira 2008). The mean growth rate of carbon dioxide in the Earth's atmosphere each year reached 2.2 ppm in 2007, up from the average of 2.0 ppm during 2000–7, and well above the 2006 rate of 1.8 ppm, leaving the concentration of carbon dioxide in the atmosphere at its highest in at least 650,000 years – and probably the highest in the preceding twenty million years (Global Carbon Project 2008). The Global Carbon Project (2008) reports that since 2000 carbon-dioxide emissions have been growing at four times the pace of the 1990s and one-third more quickly than during the preceding two decades. These changes signify a carbon cycle that is now causing 'stronger climate forcing and sooner than expected' (Global Carbon Project 2008). Indeed, carbon-dioxide emissions have been growing at about 3.5 per-cent per year since 2000, compared to less than 1 per-cent per year during the 1990s (Raupach et al. 2007: 10288). Growth in emissions from 2000 to 2007 went beyond even the *highest* forecasts of the Intergovernmental Panel on Climate Change, meaning that current trends in emissions exceed the panel's *worst-case* scenario (Global Carbon Project 2008).

Richard Perkins (2008: 46) observes that between 1990 and 2005 energy-related carbon-dioxide emissions in developing countries increased by 86 per cent, compared to an increase of 16 per cent in developed countries, with estimates showing that 'approximately three-quarters of the increase in global CO_2 [carbon dioxide] emissions up to 2030 will take place in developing countries [meaning] that, even if developed countries were to make deep emissions cuts (60 to 80 percent), the goal of avoiding dangerous climate may still elude the international community' (Perkins 2008: 46). While the United States continued to increase its carbon emissions (by about 2 per cent each year), the bulk of increases from 2006 to 2007 came from developing countries, particularly China and India, continuing the trend from 1990 to 2005. China alone accounted for fully half of the global increase in 2007 as its emission went up by 7.5 per cent (Associated Press 2008). With China now the largest national source of greenhouse gas emissions, things look set to grow worse as India is expected to overtake Russia as the third largest emitter (Associated Press 2008). Consequently, predictions of the Intergovernmental Panel on Climate Change that global temperatures will increase by 1.4 to 5.8 degrees Celsius by 2100 (IPCC 2007a) is likely to be *far too optimistic*.

All of this means that, although the 2007 fourth assessment report of the intergovernmental panel is now widely accepted as the authoritative word on climate change causes and impacts, it is clear that the panel has been 'underestimating the risks of adverse impacts . . . and the impacts previously considered to be at the upper end of likelihood are now more probable' (Climate Institute 2007: 2). Indeed, additional scientific reports, including many released after the panel's 2007 assessment, show that global warming is accelerating at rates far beyond those forecast by research that fed into the panel's report (Tin 2008: 2). One manifestation of this is a much faster deterioration of summer Arctic sea ice, which is disappearing three decades sooner than expected. This could contribute to *abrupt* climate change, instead of slow, gradual warming, by adding to a positive feedback loop as Arctic waters absorb solar energy during summer rather than reflect it back into space, in turn reducing ice thickness even more and accelerating melting each summer, as well as affecting ocean circulation that could have much wider impacts on climate. Some scientists are now saying that the Arctic Ocean could be free of ice during summer sometime between 2013 and 2040 – for the first time in over one million years (Tin 2008: 3). Sea-level rise this century could be more than double the worst estimate of the inter-governmental panel, with a possible 1.2–metre rise putting vast coastal areas around the world at risk for catastrophic impacts. What is more, the ability of natural sinks, such as the land, forests and seas, to absorb carbon dioxide is declining, meaning that more carbon dioxide will remain in the atmosphere and thus add yet more to global warming (Tin 2008: 2).

Keeping global average temperatures to below an increase of 2 degrees Celsius – a commonly stated objective of governments to avoid the most dangerous impacts from climate change – would require cuts in global greenhouse gas emissions of at least 80 percent by 2050 (stabilis-ing atmospheric concentrations of greenhouse gases at about 400–70 ppm), but even this would not avoid significant adverse impacts of climate change (Tin 2008: 2). However, according to the International Energy Agency (2006: 78), global carbon dioxide emissions will be half again larger than current emissions by 2030. This is a far cry from *cutting* global emissions, let alone cutting them by the amounts required. According to Robert Watson, former chairman of the Intergovern-mental Panel on Climate Change, if the world continues on its current emissions path until mid-century, emissions from developed countries 'would go up by 60 per cent and developing countries by about 140 per cent. If you want to hit something like 500 or 550 [ppm, which we should note is *far* higher than the level deemed safe], the OECD [Organisation for Economic Cooperation and Development] countries, instead of going up by 60 per cent, would have to go down by 60 per cent, and the developing countries, instead of going up by 140 per cent, would go up only 60 per cent – significant changes' (quoted in Rafferty 2008). To keep global temperatures to no more than a 2-degree Celsius increase over historical levels would require emitting ten times less carbon dioxide than would arise with business as usual through this century: 'You would have to have per capita emissions much more like devel-oping countries are today for everybody rather than like the US numbers' (Watson, quoted in Rafferty 2008).

James Hansen et al. (2008) have shown that, because of the time lag before the full impact of emissions is felt, even *current* concentrations of carbon dioxide in the atmosphere will probably bring the dangerous interference to the Earth's climate system that the climate change agreements were intended to prevent. The current concentration of carbon dioxide in the atmosphere – about 385 ppm – is 'already too high to maintain the climate to which humanity, wildlife, and the rest of the biosphere are adapted' (Hansen et al. 2008: 15). Even the relatively ambitious aim of the European Union to keep global temperatures to only 2 degrees Celsius above pre-industrial levels is far too weak a target. Instead, what is required as a minimum is an effort to bring carbon-dioxide concentrations down, very quickly, to about 350 ppm, meaning a near-total move away from the use of fossil fuels if carbon cannot be captured and permanently stored – a goal that is not practically (or politically) feasible at present (if ever). According to Hansen et al. (2008: 15), 'present policies, with continued construction of coal-fired power plants without CO_2 [carbon-dioxide] capture, suggest that decision-makers do not appreciated the gravity of the situation. We must begin to move now toward the era beyond fossil fuels. Continued greenhouse gas emissions, for just another decade, practically eliminate the possibility of near-term return of atmospheric composition beneath the tipping level for catastrophic effects.' Incredibly, even this pessimistic assessment is very likely to be overly optimistic. According to Kevin Anderson and Alice Bows (2008: 18), it is extremely unlikely that governments will be able to achieve greenhouse gas cuts necessary to achieve stabilisation at 450 ppm, let alone 350, with even 650 ppm considered a highly improbable objective that would require 'unprecedented rates of decarbonization' (exceeding 6 percent annually) in developed countries.

Consequently, the Kyoto Protocol is, at best, the tiniest of steps towards greater action. Indeed, the protocol, even if fully implemented as initially planned (meaning more cuts than have actually been agreed to be implemented), would delay global warming barely at all, putting off warming that would occur anyway by the end of this century by only half a dozen years or so (Dobson 2005: 265). In the meantime, global greenhouse gas emissions will continue to rise precipitously, notably because large developing countries will be increasing their use of fossil fuels as their economies grow. Climate change will continue, virtually unabated, short of new, *much more aggressive* collective action to reduce greenhouse gas emissions. However, strong signals of the much more robust action that is needed are distinct in their absence. Although some observers believe that climate diplomacy and the agreements that have resulted are remarkable and important steps towards the kind of action that scientists say is required, it is perhaps more accurate to describe those steps – relative to the scale of the problem and its anticipated impacts on people, communities and other life on Earth – as the proverbial drop in the ocean. This is not to say that there is no movement in the right direction or that a few countries are not bearing a substantial burden from efforts to reduce their greenhouse gas emissions, but *very* little has been done given what is required. Even as ideas about the threat from climate change (that is, 'climate security') and its ethical implications (for example, 'climate justice') have taken hold and started to shape politics domestically and internationally, the response has been far too modest. As such, the response to climate change reveals the most profound tragedy of the commons ever experienced.

The developing countries, despite their far larger populations, have contributed only about 20 per cent of cumulative historical carbon-dioxide emissions since 1751, but since 2005 developing countries have produced more than half (53 per cent in 2007) of global emissions (Global Carbon Project 2008). While this does not change the ethical calculus – the wealthy countries and wealthy people are the primary causes of climate change pollution – it does change the practical calculus. The Intergovernmental Panel on Climate Change, in a typically guarded understatement, characterises the failure of the Kyoto Protocol: 'To be more environmentally effective, future mitigation efforts would need to achieve deeper reductions [than the protocol] covering a higher share of global emissions' (IPCC 2007a: 62). Indeed, international legal instruments intended to avert dangerous interference with the Earth's climate – the stated aim of the climate change convention – are increasingly about mitigating and adapting to that dangerous interference, rather than averting it, in part because perceptions of international (in)justice provide a poor guide for getting at the growing sources of the problem (addressed in the next chapter) – even, one might argue, giving developing countries that are capable of acting licence to do nothing while waiting for the developed countries to do all that they should. Those developed countries see little point in acting, at least in practical terms, if the developing states hold back. The result is the current tragedy of the atmospheric commons.

A primary argument used by all governments, and one that garners support from their constituents, is that the costs of addressing climate change are just too great. However, Nicholas Stern (2007) has shown that the long-term costs of not limiting greenhouse gas emissions will be greater than the costs of not doing so. William Nordhaus (2008: 195) has shown that 'an ideal and efficient' set of climate change policies entailing gradually increasing emissions cuts over time and set to 'maximise economic welfare of humans' would be 'relatively inexpensive and would have substantial impact on long-run climate change'. The details of proposed plans that save long-term costs by acting sooner, or at least begin the process of acting and ramping it up with time, vary, but the general message is clear: the costs are relatively low, even by purely economic measures, when compared to not acting. Nevertheless, many entrenched interests, including many governments and industries, remain worried about the costs of action. This prevents it from happening. But, even if we did not know (as we do) that action would cost less than inaction, there is a 'viciousness in the thought that the cost of mitigation and adaptation should be a reason for doing little or nothing. It amounts to harming others for money' (Garvey 2008: 111). So it is doubly wrong for the affluent of the world not to act with great urgency.

Aaron Maltais (2008: 599) has described the tragedy of the atmospheric commons, at least with respect to mitigation, this way: 'Because it is the total global reductions in GHG [greenhouse gas] emissions (and not where these reductions take place) that has an effect on climate conditions, individuals, companies, and states all have self-interested reasons to free-ride by letting others take on the costs of mitigation while they themselves continue to enjoy the benefits of those activities that cause atmospheric pollution.' When people, businesses and government put this reasoning into action, 'they will collectively fail to adequately provide the public good and/or fail to protect the "common pool resource" of climate security' (Maltais 2008: 599). As Speth (2008: 71) puts it, 'the results of two decades of international environmental diplomacy are deeply disappointing. The bottom line is that today's treaties and their associated agreements and protocols cannot drive the changes needed.' This is because those treaties are too (intentionally) weak and easy for governments to ignore. Even the response of the European Union, which attempts to bridge the gap between the United States and the developing world by 'going first' to some extent, is very small relative to environmental and human needs. 'In sum,' Speth (2008: 72) laments, 'global environmental problems have gone from bad to worse [and] governments are not prepared to deal with them . . .'.

Benito Muller (2002) has shown how there is a 'great divide' between conceptions of climate justice ('equity') in developed countries and conceptions in developing countries, and that the former's conception ('mitigation myopia' (Muller 2002: 40)) has been most pronounced in the climate change regime. According to Muller (2002: 1), in developed countries 'the paramount climate change equity problem . . . is regarded to be the issue of allocating emission mitigation targets; in the South, the concern . . . is above all about the discrepancy between the responsibility for, and the sharing of, climate impact burdens'. For developing country stakeholders (that is, governments) in climate change negotiations, the most pressing matter of justice is 'having to bear human impact burdens disproportionate with causal responsibilities', a matter that they believe has been ignored in negotiations (Muller 2002: 1). What is more, Muller (2002: 1) argues that, while justice has been made part of the climate change regime as a consequence of pressure from developing countries, the scope of justice – 'namely emission mitigation – has been firmly set by the industrialized countries'. He believes that the explanation for these differing perspectives on climate justice can be found in how developed and developing countries view the problem of climate change. The former see it as an ecological problem where the 'overriding moral purpose' is environmental integrity, with injustice an issue insofar as it becomes an obstacle to realising that purpose. In contrast, for the developing countries, 'climate change has primarily come to be seen as a human welfare problem' in which the victim is 'not "Nature", but people, and the paramount inequity is one of human victims and human culprits' (Muller 2002: 2).

CONCLUSION

The political world is made up of sovereign states, for better or worse, so it is perhaps normal for most discussions of climate justice to be about national communities vis-à-vis one another. According to the dominant 'morality of states', states have rights and bear the burden of not violating other states' rights. Thus to say that affluent states have an obligation to reduce their greenhouse gas emissions and to aid poor countries that will suffer from climate change is not profound. As states go, the developed ones have caused most of the problem and the developing ones are those that will suffer the most from it. Common conceptions of fairness demand that the former act first and aid the latter accordingly. The climate change convention and associated agreements have affirmed this. States have recognised the importance of international justice in this context, but they have consistently failed to implement it.

The response to climate change has been premised on the rights of states – their rights to choose voluntarily the degree to which they will participate (or not) in collective international action to address the problem. Insofar as justice has been part of the climate change regime – and indeed it has – it has been premised additionally on the duties of states. But the rights and duties of *individuals* have played little part, except insofar as individual rights of people within particular countries have been used by governments to bolster their arguments for or against taking on state obligations to address climate change. As Simon Caney (2005a: 773) stresses, the principle of common but differentiated responsibility 'refers to the responsibilities of *states*' (emphasis added). Elliott (2006: 357) observes that common but differentiated responsibility has been focused on 'managing the political relationships between rich and poor countries, [while] much less attention has been paid to the issue of compensatory justice for poor people. In other words, it is not individuals who are being considered here. This runs the risk of reducing burden sharing to a form of proxy cosmopolitanism – confining justice to the inter-state level.'

Hermann Ott and Wolfgang Sachs (2002) point out that states have been made the locus of responsibility in the context of climate change: 'This puts climate policy firmly into the framework of what can be called "the Westphalian constellation". In this framework, the world of nation-states . . . was seen as a series of containers which hold a society and all its layers within a territorially bounded space. As these containers burst open with globalisation, some of the "Westphalian" assumptions become more and more fictitious' (Ott and Sachs 2002: 166). Far from being a solution to environmental problems, international justice in the context of climate change has been at best a justification for giving developing states a bit more aid, which they deserve, but has if anything been a kind of 'curse' that has preoccupied diplomats and prevented them from seriously and fully discussing the role of *people* per se as causes of climate change, thus avoiding where the source of the problem really lies – with the people who actually cause the most pollution and are capable of reducing it.

There is no shortage of international agreements on all manner of environmental issues, including climate change. But, as Dower (2007: 172) puts it, 'still many people sense that we have not really changed tack in any major way'. The climate change regime may do something similar to the long-standing international trade regime: the focus on national obligations, technology transfers and even the growing number of international climate funds tends to empower national elites and the well-off while saying nothing about how they ought to behave themselves. The result is much talk of addressing the problem, but little action intended primarily for that purpose. And those who benefit the most from the status quo – rich people everywhere – continue to win, usually at the expense of the poor.

The developing countries have overtaken the developed ones in their national emissions of greenhouse gases. Thus it is essential that the large developing countries eventually limit their greenhouse gases if climate change is to be minimised. The question becomes how this can happen when the climate change regime is premised on those countries waiting for major emissions cuts in, and aid from, the developed states. By putting international justice at the centre of the climate change regime, it became fairer and thus participation became more comprehensive. But this has not resulted in anything like the kind of action that is required to address both the levels of pollution going into the atmosphere and the consequences for those who will suffer the consequences. What is fair and just from the perspective of international justice is not necessarily fair and just from other perspectives. It can be the opposite. To be sure, it would not be fair if China and other less-developed (least of all very poor) countries were required to take on the same obligations to combat climate as the United States and other affluent countries. But it is also not fair, nor is it environmentally sound, for the many affluent people in developing countries, and especially rich elites there, to be absolved of duties regarding climate change. Why, ethically, should a poor person in, say, Germany be lumped with the wealthy of Germany to protect and aid both the poor *and* the rich in China or other developing countries, especially when the latter may pollute far more? The belief that affluent countries ought to aid poor ones in the context of climate change, and that the former ought to be drastically cutting their greenhouse gas emissions while allowing the latter to increase theirs, seems grossly inadequate, even though it is ethically well grounded from an interstate perspective.

The failures of a climate change regime premised on prevailing interstate doctrine suggest that an alternative ethic is needed. If that ethic is not to replace the doctrine of international environmental justice, it must at least improve upon it. As Part III argues, cosmopolitan ethics and global justice are much more suited to the realities of climate change, modernity and globalisation than are extant forms of justice premised on the morality of states. Global justice brings people more explicitly into the climate change regime while also helping to create conditions in which international climate justice is more likely to be actualised.

NOTES

1. The developing countries are in fact a disparate group with different views. For example, Saudi Arabia and other oil-producing states do not want to see the use of petroleum reduced, but Maritius and other small-island states very much want that to happen (see Muller 2002).
2. The principle of common but differentiated responsibility here applies to all countries, not just the developed–developing-country relationship. Consequently, some developed countries have more responsibility than other developed countries. This differentiation is based ostensibly on national circumstances, but in reality it was largely a function of political bargaining.

There is also the important question of how future emissions limitations should be distributed among developing countries themselves. Some of the newly industrialised countries and those with substantial wealth, such as Saudi Arabia, Singapore and South Korea, as well as rapidly industrialising countries, such as Brazil and China, will presumably be expected to act long before the very poorest countries are expected to do anything more than receive aid for mitigation of their suffering from climate change (cf. Ott et al. 2005: 91).

3. Jagers and Duus-Otterstrom (2008) argue that the assignment of responsibilities depends on whether burdens are distributed for mitigation or for adaptation. They critique the more common approach of looking at mitigation and adaptation as a 'package deal' à la Caney (2005a) and Page (2006), and which is generally the approach here.

PART III
GLOBAL JUSTICE

CHAPTER 5

COSMOPOLITAN ETHICS AND JUSTICE

Governments have agreed to about 200 treaties dealing with environmental and natural resource issues. Some of these treaties have met with success in limiting and even reversing pollution and overuse of resources. However, one consequence of the communitarian Westphalian norms and the international doctrine upon which these agreements have been based is that states have usually not arrived at *effective* means to protect natural resources and the environment when doing so might undermine states' perceived national interests. In most cases, a tragedy of the commons obtains: it is usually in the interests of states to allow continued exploitation of environmental commons and resources because, without effective enforcement, doing otherwise leaves open the likelihood that other countries will exploit the resources for themselves, leaving those who act in the common interest relatively worse off. Consequently, the focus on the interests of states, which is an extension of the Westphalian norms that served many states relatively well until recent decades, has in the case of major environmental problems – and profoundly in the case of climate change – been a sort of curse on often well-meaning aims of governments and other actors to tackle environmental problems.

As Chapter 4 showed, the urgency of addressing climate change is much greater than anticipated by most scientists just a few years ago (as described in Chapter 1). From whence should come the arguments for dealing with this? As Nigel Dower (2007: 184) argues, 'there is an urgency about creating much stronger norms both in the cultures of communities and [in] the working practices of states. These cultures and practices are vital, but the arguments for creating them come from, and must come from, elsewhere.' One potentially potent remedy to the 'curse of Westphalia' that characterises today's response to climate change can be found in cosmopolitan ethics and global conceptions of justice that focus on *people* as well as on states. Cosmopolitans recognise the rights, obligations and duties of capable individuals regardless of nationality; they see American and Chinese *people*, for example, rather than the United States or China as *states*. Cosmopolitanism attributes duties to capable persons for a number of reasons, such as Henry Shue's (1995) principle of 'do no harm', Peter Singer's (2003) historical argument of 'you broke it, you fix it' and his maxim to 'prevent extreme suffering'; Dale Jamieson's (1997) belief in the 'ability to benefit others or prevent harm'; Brian Barry's (1998) 'priority of vital interests'; and Onora O'Neill's (1988) Kantian prescription to avoid undermining the ability of others to be independent moral agents. *International* justice considers national borders to be both the practical and the ethical foundation for justice. But cosmopolitan or *global* justice, while recognising that national borders have practical importance, views them as the wrong (or at least an inadequate) basis for deciding what is just – including in the context of climate change. Climate change creates the conditions for cosmopolitan responses, in large part by way of the interdependence it engenders.

This chapter looks at cosmopolitanism in more detail. It introduces several scholars' definitions of this world ethic, noting many of its key features. The chapter then introduces and critiques some cosmopolitan (or cosmopolitan-like) arguments about climate change, in the process showing how cosmopolitanism is suited to a better understanding of the problem of climate change. As we will see, while cosmopolitanism can help to overcome the state-centric myopia of the climate change regime, even most cosmopolitan prescriptions tend to revert to discourse and policy related to states. The next chapter explores why state-centric thinking and responses are problematic in the extreme, and shows the relative utility of cosmopolitanism. Chapter 7 then proposes a partial alternative to international doctrine, one premised on cosmopolitan ethics and the practical role of people, although even that prescription must account for the enduring role of states.

COSMOPOLITANISM AND GLOBAL JUSTICE

In contrast to the Westphalian norms that have guided and indeed defined the international system for centuries – a set of ethics premised on protecting the interests of the state – cosmopolitans envision an alternative way of ordering the world. Cosmopolitans want to 'disclose the ethical, cultural, and legal basis of political order in a world where political communities and states matter, but not only and exclusively. In circumstances where the trajectories of each and every country are tightly entwined, the partiality, one-sidedness and limitedness of "reasons of state" need to be recognised' (Held 2005: 10). As Gillian Brock and Harry Brighouse (2005: 7) argue, 'it is only when foreign policy is liberated from its focus on exclusively furthering the national interest that we can begin to ask the right questions', such as how to balance concerns for human rights everywhere with the narrower interests of our national compatriots or of the state.

FEATURES OF COSMOPOLITANISM

Cosmopolitans have included the ancient Cynics, like Diogenes, Stoics like Cicero and Chinese philosophers like Mo Tsu. Enlightenment cosmopolitans included Voltaire, Jeremy Bentham and Immanuel Kant. Contemporary cosmopolitans include Brian Barry, Charles Beitz, Simon Caney, Nigel Dower, Martha Nussbaum, Peter Singer and an expanding list of thinkers (see Caney 2005b: 4). Thomas Pogge (2008: 175) sums up three core elements of cosmopolitanism this way:

> First, *individualism*: the ultimate units of concern are *human beings*, or *persons* – rather than, say, family lines, tribes, ethnic, cultural, or religious communities, nations, or states. The latter may be units of concern only indirectly, in virtue of their individual members or citizens. Second, *universality*: the status of ultimate unit of concern attaches to *every* living human being *equally* – not merely to some sub-set, such as men, aristocrats, Aryans, whites, or Muslims. Third, *generality*: this special status has global force. Persons are ultimate units of concern for *everyone* – not only for their compatriots, fellow religionists, or such like.

David Held (2005: 12) has synthesised cosmopolitanism into a set of eight, universally shared, key principles: '(1) equal worth and dignity; (2) active agency; (3) personal responsibility and accountability; (4) consent; (5) collective decision-making about public matters through voting procedures; (6) inclusiveness and solidarity; (7) avoidance of serious harm; and (8) sustainability.' From these principles the 'cosmopolitan orientation' emerges: 'that each person is a subject of equal moral concern; that each person is capable of acting autonomously with respect to the range of choices before them; and that, in deciding how to act or which institutions to create, claims of each person affected should be taken equally into account' (Held 2005: 15). Importantly for climate change, the last two principles provide 'a framework for prioritising urgent need and resource conservation. By distinguishing vital from non-vital needs, principle 7 creates an unambiguous starting point and guiding orientation for public decisions [and] clearly creates a moral framework for focusing public policy on those who are most vulnerable' (Held 2005: 15). A 'prudential orientation' is set down by principle 8 'to ensure that public policy is consistent with global ecological balances and that it does not destroy irreplaceable and non-substitutable resources' (Held 2005: 15–16).

Some cosmopolitans take a consequentialist perspective, such as Singer's (1996) utilitarianism, while others take a deontological perspective, such as Caney's (2005a) global political theory premised on human rights. Charles Jones (1999: 15) describes three 'species' of cosmopolitanism: utilitarianism, human rights and Kantian ethics. He defines cosmopolitanism as a moral standpoint that is 'impartial, universal, individualist, and egalitarian. The fundamental idea is that each person affected by an institutional arrangement should be given equal consideration. Individuals are the basic units of moral concern, and the interests of individuals should be taken into account by the adoption of an impartial standpoint for evaluation' (Jones 1999: 15). The nature of cosmopolitanism might be best appreciated by pointing to what it rules out: 'it rules out the assigning of ultimate rather than derivative value to collective entities like nations or states, and it also rules out positions that attach no moral value to some people, or weights the value people have differently according to characteristics such as ethnicity, race, or nationality' (Brock and Moellendorf 2005: 2). Another way of looking at cosmopolitanism, particularly in practice, is that it 'does not privilege the interests of insiders over outsiders' (Linklater 2001: 264; cited in Elliott 2006: 345). In a fundamental way, what is crucial about the cosmopolitan perspective is its 'refusal to regard existing political structures as the source of ultimate value' (Brown 1992: 24, cited in Barry 1999: 36).

Two versions of cosmopolitanism are routinely identified: an ethical/moral/normative version, which focuses on the underlying moral arguments regarding how people, states and other actors should justify their actions in the world, and an institutional/legal/practical version, which aims to translate ethics into institutions and policies. Pogge (1992, 2008) distinguishes between moral and legal cosmopolitanism. Moral cosmopolitanism points to the moral relations among people; 'we are required to respect one another's status as ultimate units of moral concern – a requirement that imposes limits on our conduct and, in particular, on our efforts to construct institutional schemes' (Pogge 2008: 175). Legal cosmopolitanism goes a step further by advocating creating institutions of global order, possibly in the form of a 'universal republic' in which 'all persons have equivalent legal rights and duties' (Pogge 2008: 175). This latter position may seem to be a bit extreme; moral cosmopolitanism certainly does not require institutionalisation of a universal republic (or 'world government'). One variant of institutional cosmopolitanism asserts that 'the world's political structure should be reshaped so that states and other political units are brought under the authority of supranational agencies of some kind' (Beitz 1999: 287, cited in Dower 2003: 27). Institutional cosmopolitans sometimes call for major, even radical, changes to global institutions, but moral cosmopolitans frequently do not see this as being necessary.

An alternative (more realistic) version of institutional cosmopolitanism 'postulates fundamental principles of justice for an assessment of institutionalised global ground rules [while also being] compatible with a system of dispersed political sovereignty that falls short of a world state' (Forst 2001: 164). As Caney (2005b: 5) points out, some moral cosmopolitans 'reject a world state. They think that cosmopolitan moral claims are compatible with, or even require, states or some alternative to global political institutions'. Thus it is entirely possible and appropriate to advocate institutions well short of world government that contribute to global order generally, and particularly global justice within specific issue areas. What is more, as Darrel Moellendorf (2002: 172) reminds us, 'very few people who have thought about these matters [i.e. whether an egalitarian world order would contain multiple states or a world-state] have considered the latter a real possibility, and with good reason' – not least the practicality of governing the world's many billions of people and the threat such a world state might pose to human rights. However, it is also clear 'that the establishment and maintenance of justice requires a significant re-conceptualization of the principle of state sovereignty [and] a coordinated international response' (Moellendorf 2002: 172).

Cosmopolitanism includes two additional features, according to Brock and Brighouse (2005: 2): identity and responsibility. The former refers, for example, to a person who is influenced by a variety of cultures or perhaps one who identifies with broader humanity rather than to a particular group or nation. The latter 'guides the individual outwards from obvious, local, obligations, and prohibits those obligations from crowding out obligations to distant others . . . It highlights the obligations we have to those whom we do not know, and with whom we are not intimate, but whose lives touch ours sufficiently that what we do can affect them' (Brock and Brighouse 2005: 3). According to Robin Attfield (2003: 162):

> Cosmopolitan ethicists maintain that ethical responsibilities apply everywhere and to all moral agents capable of shouldering them, and not only to members of one or another tradition or community, and that factors which provide reasons for action for any agent, whether individual or corporate, provide reasons for like action for any other agent who is similarly placed, whatever their community may be or believe. They also deny limits such as community boundaries to the scope of

responsibilities; responsibilities (they hold) do not dwindle because of spatial or temporal distance, or in the absence of reasons transcending particular facts or identities.

One might also think of both weak and strong forms of cosmopolitanism, the former saying that some obligations obtain beyond the society or the state, while the latter says that any principles (of justice, for example) that apply within the state also apply worldwide. As Brock and Brighouse (2005: 3) see it, 'everyone has to be at least a weak cosmopolitan now if they are to maintain a defensible view, that is to say, it is hard to see how one can reject a view that all societies have *some* global responsibilities'. Pogge (2002a: 86) addresses critics of 'weak' cosmopolitanism – 'the anodyne view that all human beings are of equal worth', which almost everyone, except 'a few racists and other bigots', accepts – and 'strong' cosmopolitanism – 'the view that all human agents ought to treat all others equally and, in particular, have no more, or less, reason to help any one needy person than any other', which it might be argued is falsely expansive – by proposing an 'intermediate' view of cosmopolitanism based on *negative* duties. From this viewpoint, the fact that someone is a fellow national citizen 'makes no difference to our most important negative duties' (Pogge 2002a: 87): 'You do not have more moral reason not to murder a compatriot than you have not to murder a foreigner. And you do not moderate your condemnation of a rapist when you learn that his victim was not his compatriot' (Pogge 2002a: 87). Intermediate cosmopolitanism 'asserts the fundamental negative duty of justice as one that every human being owes to every other' (Pogge 2002a: 89). But, just as duties of justice vary *within* communities – it is widely accepted that one can have a greater duty to family members than to the wider community – this does not mean that there are no duties whatsoever, in particular no duty to avoid contributing to conditions that undermine the fundamental rights and needs of others within the community. Similarly, while we may favour compatriots in many ways, we ought not to support institutions that impose an unjust order on people living in other communities. According to Pogge (2002a: 90–1), 'special relationships can *increase* what we owe our associates, but cannot *decrease* what we owe everyone else'. The upshot is that, 'though we owe foreigners less than compatriots, we owe them something. We owe them negative duties, undiluted' (Pogge 2002a: 91).

For cosmopolitans, 'the world is one domain in which there are some universal values and global responsibilities' (Dower 2007: 28). Cosmopolitan responsibility entails 'the recognition that since we live, in some sense, in one global community or society – whether or not most of us have much of a feeling for this – we do have duties to care in one way or another about what happens elsewhere in the world and to take action where appropriate' (Dower 2007: 11). It is not enough to identify with humanity to be a cosmopolitan; it is necessary to act (or to be willing to act) accordingly. From this basis, it stands to reason that capable individuals are obliged to act even if they live in dissimilar communities (that is, rich or poor countries), and those who are more capable are more responsible to do so. James Garvey puts it this way: 'the better placed an individual is to do what is right, the greater the onus on him to do what is right' (Garvey 2008: 82).

Cosmopolitans frequently justify their claim that justice ought to prevail globally by using one or both of two arguments (cf. Forst 2001: 163). One argument, sometimes building on John Rawls's (1971) domestic theory of justice, is that levels of international cooperation today are extensive enough to make international society sufficiently like domestic society to warrant applying justice principles that were previously the domain of domestic communities to world affairs (see Beitz 1979b).[1] Another argument, derived from the empirical realities of globalisation and the interdependencies and cause-and-effect relationships it manifests, is that justice ought to prevail globally because people and communities, whether knowingly or not, intentionally or not, increasingly affect one another, sometimes in profound ways. Justice is demanded by this latter argument because globalisation is in large part a process of redistribution of scarce resources away from those with the least to those with the most. David Weinstock (2007: p. ix) describes a relatively new 'way of understanding the relationship between the global rich and the global poor[:] the fate of the global rich is not as causally independent of the plight of the global poor as had previously been thought . . . According to this view, globalization makes it the case that our obligations toward the global poor are obligations of *justice* rather than of *charity* . . .'. Climate change could be the most profound manifestation of this latter argument.

COSMOPOLITAN JUSTICE

Most cosmopolitans accept, and often advocate, duties of *global* justice for states and frequently for individuals. Global justice is based upon a cosmopolitan world ethic premised on the rights, duties and ethical importance – and moral pre-eminence – of persons. According to Dower (2007: 92), the wish for global justice is motivated by three claims: (1) 'obligations are substantial or significant, rather than minimal or merely "charity" '; (2) global obligations should be premised on 'institutional arrangements which specify quite clearly which bodies have which duties to deliver justice'; and (3) obligations have their foundation in 'the human rights of others which are either violated by the global economic system or fail to be realised because of it'. For cosmopolitans, 'the world is a community of people and not a set of countries: that is, it is a community in which all have a claim to justice, just as they themselves owe justice to others' (Sachs and Santarius 2007: 125).

Onora O'Neill (2000: 191) proposes a practical approach to determining who has moral standing: 'Questions about standing can be posed as context-specific *practical* questions, rather than as demands for comprehensive theoretical demarcations.' Answers are found in part in the assumption that people 'are already building into our action, habits, practices and institutions' (O'Neill 2000: 192). This suggests a 'more or less cosmopolitan' approach to principles of justice in given contexts (O'Neill 2000: 192). O'Neill's practical approach offers a *relational* account of moral standing:

> Conjoined with the commonplace facts of action-at-a-distance in our present social world, this relational view points us to a *contingently* more or less cosmopolitan account of the proper scope of moral concern in some contexts. We assume that others are agents and subjects as soon as we act, or are involved in practices, or adopt policies or establish institutions in which we rely on assumptions about other's capacities to act and to experience and suffer. Today we constantly assume that countless others who are strange and distant can produce and consume, trade and negotiate . . . pollute and or protect the environment . . . Hence, *if* we owe justice (or other forms of moral concern) to all whose capacities to act, experience and suffer we take for granted in acting, we will owe it to strangers as well as to familiars, and to distant strangers as well as to those who are near at hand . . . Today only those few who genuinely live the hermit life can consistently view the scope of moral concern which they must acknowledge in acting as anything but broad, and in some contexts more or less cosmopolitan. (O'Neill 2000: 195–6)

This is a view of justice that takes *obligation* as being essential; 'obligations provide the more coherent and more comprehensive starting point for thinking about . . . the requirements of justice' than do rights because it is hard to know who has harmed which distant others (O'Neill 2000: 199). Dower (2007: 178) must be right when he says that 'most thinkers would adopt the following maxim: where the lines of cause and effect run across nation-state borders, so do the lines of moral responsibility. To accept such a maxim is implicitly to endorse a "global ethic", according to which the whole world is one moral domain, and the network of moral relationships extends in principle across the world.'

Andrew Dobson (2006) makes a case for cosmopolitan obligation arising from the causal impacts of globalisation in its many manifestations, including global environmental change. What is especially important about his argument is that he goes beyond cosmopolitan morality and sentiment, which are important but apparently not sufficient to push enough people to act. Dobson describes 'thick cosmopolitanism', in particular the source of obligation for cosmopolitanism, in an attempt to identify what will motivate people (and other actors) not only to accept cosmopolitanism but to act accordingly. While he seems to accept that we are all members of a common humanity, he is unhappy with leaving things there: 'Recognising the similarity in others of a common humanity might be enough to undergird the principles of cosmopolitanism, to get us to "be" cosmopolitans (principles), but it doesn't seem to be enough to motivate us to "be" cosmopolitan (political action)' (Dobson 2006: 169). Common humanity is one basis for cosmopolitanism, but it does not create the 'thick' ties between people that arise from causal responsibility. Dobson argues that the way to think about the 'motivational problem is in terms of nearness and distance . . . to overcome the "tyranny of distance"' (Dobson 2006: 170). He invokes Linklater's suggestion that, if we are 'causally responsible for harming others and their physical environment', we are far more likely to act as cosmopolitans should (Linklater 2006: 3, quoted in Dobson 2006: 172). Relationships of causal responsibility 'trigger stronger senses of obligation than higher-level ethical appeals can do' (Dobson 2006: 182).

To help us comprehend this connection between nearness, causality and motivation, Dobson describes a Good Samaritan whose actions to assist a suffering man move us because the Samaritan was not responsible for the man's injuries; the Good Samaritan acted purely out of beneficence. However, if the Samaritan were 'implicated in the man's suffering in one way or other, we would *expect* him to go to his aid and his act of succour would seem less remarkable' (Dobson 2006: 171). This illustrates the 'cosmopolitan nearness' that arises from causal responsibility (Dobson 2006: 172–3). While this might not be the whole story – we might have obligations to help others in critical need simply because we are *capable* of helping them – even if we do have obligations for other reasons they are amplified if we are indeed the cause of the harm in question. What is more, in keeping with cosmopolitan morality, the 'causal responsibility approach' that Dobson describes is universal; 'the obligation to do justice implicit in it is owed, in principle, to absolutely everyone without fear or favour' (Dobson 2006: 173). It is also universal because we now live in a globalised world in which most of what we, the affluent people of the world, do involves relations of causal responsibility, therefore making them relations of justice. As Dobson (2006: 178) puts it, 'the ties that bind are not, therefore, best conceived in terms of the skin skein of common humanity, but of chains of cause and effect that prompt obligations of justice rather than sympathy, pity, or beneficence'. As O'Neill (2000: 187) argues, 'in our world, action and inaction at a distance are possible. Huge numbers of distant strangers may be benefited or harmed, even sustained or destroyed, by our action, and especially by our institutionally embodied action, or inaction – as we may be by theirs'. Dobson's (2005: 270) point is that, in these kinds of relations of actual harm, justice 'is a more binding and less paternalistic source and form of obligation than charity'.

The cosmopolitan standpoint presents serious challenges to the prevailing climate change regime, but it also offers opportunities for new, possibly more effective prescriptions for making the regime more attuned to reality and thus for making it more effective. When it comes to justice, as Caney (2005a: 749) points out, 'whereas conventional theories of distributive justice concern themselves with the distribution of burdens and benefits within a *state*, the issues surrounding climate change require us to examine the *global* distribution of burdens and benefits'. This means that 'thinking about justice – and particularly of environmental justice as an integral component of global justice – can no longer be confined within the boundaries of nation-states' (Hayden 2005: 133) if we want to comprehend the realities of climate change.

What more do cosmopolitans say about climate change?[2] The next section scratches the surface of their arguments, setting the stage for further discussion in later chapters.

COSMOPOLITANISM AND CLIMATE CHANGE

Lorraine Elliott (2006: 350) asserts that environmental harms crossing borders 'extend the bounds of those with whom we are connected, against whom we might claim rights and to whom we owe obligations within the moral community'. She describes this as a 'cosmopolitan morality of distance', which effectively creates 'a cosmopolitan community of duties as well as rights' (Elliott 2006: 350). Elliott (2006: 351) argues that this obtains for two reasons: 'the lives of "others-beyond-borders" are shaped without their participation and consent [and] environmental harm deterritorialises (or at least transnationalises) the cosmopolitan community. In environmental terms, the bio-physical complexities of the planetary ecosystems inscribe it as a global commons of a public good, constituting humanity as an ecological community of fate.' Consequently, Elliott (2006: 363) believes that the cosmopolitan standpoint provides a better 'theoretical and ethical road map for dealing with global environmental injustice' than does international doctrine. Attfield (1999: 205) goes further, arguing that only cosmopolitanism can do 'justice to the objective importance of all agents heeding ethical reasons, insofar as they have scope for choice and control over their actions, and working towards a just and sustainable world society'.[3] He believes that criticisms of failed state responses to environmental problems will inevitably be based on cosmopolitanism because 'the selective ethics of nation states are liable to prioritise some territories, environments, and ecosystems over others. If this meant nothing but leaving the other environments alone, this might not be too pernicious. [However,] it often means not leaving alone the others but polluting or degrading them' (Attfield 2005: 41). Derek Heater (1996: 180) also critiques what he calls the 'traditional linear model of the individual having a political relationship with the world at large only via his state', because, at least if we are concerned about 'the integrity of all planetary life, the institution of the state is relegated to relative insignificance – if not, indeed, viewed as a harmful device'.

This points to the need for a theory of environmental justice that fully encompasses the causes and consequences of climate change (cf. Caney 2006a: 53). Such a theory must almost certainly be cosmopolitan, as Steve Vanderheiden (2008: 104) argues:

> Insofar as a justice community develops around issues on which peoples are interdependent and so must find defensible means of allocating scarce goods, global climate change presents a case in which the various arguments against cosmopolitan justice cease to apply. All depend on a stable climate for their well-being, all are potentially affected by the actions or policies of others, and none can fully opt out of the cooperative scheme, even if they eschew its necessary limits on action. Climate change mitigation therefore becomes an issue of cosmopolitan justice by its very nature as an essential public good . . .

Governments have agreed to some principles and practices of environmental justice that apply at the *interstate* level (Chapter 3), including in the context of the climate change regime (Chapter 4). Indeed, some of the related proposals have a cosmopolitan flavour. For example, the developing countries have called for the allocation of greenhouse gas emissions to be based on equal per-capita allotments. This would in effect require the rich countries to pay poor ones for the use of the latter's allotments. While the climate change negotiations so far have arrived at bargains that fall short of codifying these equal per-capita rights to the atmosphere, Frank Biermann (2005: 20) believes that it is only such allotments that 'have an inherent appeal due to their link to basic human rights of populations in both South and North' and that will probably have 'the normative power to grant the climate governance system the institutional stability it needs in the decades and centuries to come', not least because only equal per-capita rights can be democratically supported. Peter Baer (2002) also argues that equal per-capita emissions rights are the only ethical option, noting that this has the practical benefit of offering options for developing-country emissions limits in the future. But others have argued that 'political and economic reasons [mean that] such a proposal has no chance of being accepted by developed countries because it leads to unacceptable costs for them . . .' (Hourcade and Grubb 2000: 199). This scepticism is well justified based on the history of the climate change regime, even as these arguments show that cosmopolitan-like positions are being debated among diplomats already – although states are often the intended bearers of duties and frequently the proposed beneficiaries of associated rights.

The work of Simon Caney (2005a, 2005b, 2006a, 2006b, 2008) is particularly noteworthy for the way it looks at climate change from a cosmopolitan standpoint, in particular showing how and why climate change is unjust because it threatens human rights. As Caney (2006b: para. 2) states, 'the current consumption of fossil fuels is unjust because it generates outcomes in which people's fundamental interests are unprotected and, as such, undermines certain key rights . . . This is unjust whether those whose interests are unprotected are fellow citizens or foreigners and whether they are currently alive or are as yet not alive.' His argument proceeds this way (see Caney 2006b: para. 23): (1) persons have a right to something if it is 'weighty enough to generate duties on others'; (2) climate change jeopardises 'fundamental interests' (for example, not suffering from drought, crop failure, heatstroke, infectious diseases, flooding, enforced relocation and 'rapid, unpredictable and dramatic changes to their natural, social and economic world' (Caney 2005a: 768)); (3) the interests jeopardised are of sufficient weight to generate obligations on other persons; thus (4) 'persons have a right not to suffer from the ill-effects associated with global climate change' (Caney 2005a: 768). One advantage of Caney's argument is that it does not turn on the question of who is causing climate change.

But the question still remains: who ought to bear the burdens of addressing the problem? Caney (2005a: 769) answers that all persons 'are under the duty not to emit greenhouse gases in excess of their quota' and persons 'who exceed their quota (and/or have exceeded it since 1990) have a duty to compensate others (through mitigation or adaptation)'.[4] He concludes that 'the most advantaged have a duty either to reduce their greenhouse gas emissions in proportion to the harm resulting from [mitigation] or to address the ill-effects of climate change resulting from [adaptation] (an ability to pay principle)', with the added proviso that 'the most advantaged have a duty to construct institutions that discourage future non-compliance . . .' (Caney 2005a: 769). In Chapter 6 I argue for action by capable (and advantaged) individuals, and later (in Chapter 7) for the creation of institutional processes intended to foster it.

Michael Mason (2005: p. x) argues that there is environmental responsibility across borders because those who produce significant harm, regardless of whether they are states, are morally obliged to consider those affected by the harm, regardless of whether those harmed are co-nationals. What he argues for requires 'an appreciation of expressions of well-being not mediated by states' (Mason 2005: 12). Jamieson (2002) has pointed out that the notion that governments have duties only to one another is problematic for environmental protection. Given the nature of environmental problems and the environmental interests and actions of different individuals and organisations, 'rather than thinking about the problem of the global environment as one that involves duties of justice that obtain between states, we should instead think of it as one that involves actions and responsibilities among individuals and institutions who are related in a variety of different ways' (Jamieson 2002: 306). Consequently, the common notion of international environmental justice – obligations of states to aid one another in this context – ought to

> be supplemented by a more inclusive ecological picture of duties and obligations – one that sees people all over the world in their roles as producers, consumers, knowledge-users, and so on, connected to each other in complex webs of relationships that are generally not mediated by governments. This picture of the moral world better represents the reality of our time in which people are no longer insulated from each other by space and time. Patterns of international trade, technology, and economic development have bound us into a single community, and our moral thinking needs to change to reflect these new realities. (Jamieson 2002: 306–7)

What comes from these views is the need to correct the preoccupation with governments and states, and to focus much more on the needs, obligations *and* actions of individuals, which is what cosmopolitanism does.

COSMOPOLITANISM AND THE STATE

Cosmopolitanism can point to new ways of addressing climate change, but it is important to point out that even cosmopolitan literature tends to draw conclusions about what states should do rather than what people should do. Even cosmopolitans often lapse into Westphalian discourse when describing how to implement their ideas in the context of climate change. O'Neill (2000: 188) is one philosopher who has noted that even cosmopolitans routinely revert to statism: 'much liberal and socialist thinking, although cosmopolitan in the abstract, is statist when it comes to action and politics.' Examples abound of cosmopolitan analysis that falls back on the discourse of states. For example, in her examination of environmental conventions, Elliott (2005, 2006) does what she argues those conventions do: invoke cosmopolitanism rhetorically without seeking to put it into practice among persons. She rightly reveals the injustices of rich *people* everywhere – but mostly those in rich countries – consuming far more per capita than the world's poor people, but she bolsters this idea by reverting to data about per-capita consumption in wealthy *states* and 'industrialised economies' (Elliott 2006: 348).

James Speth (2008: p. x) asks us to consider how serious the environmental crises really is today by revealing one simple measure of the problem: to destroy the Earth's atmosphere and species, all we have to do is continue what we are doing right now – to continue polluting at current levels. But the situation is even worse than that; pollution is *not* staying at current levels – it is *increasing*. In looking for solutions to this problem, Speth does something that countless other authors have done: he focuses on 'materialism and consumerism in today's affluent societies' (Speth 2008: 10). Speth eloquently identifies the environmental consequences of affluence. However, like most of those other authors, he ignores much of the actual source of the problem: the global materialists and consumers, roughly half of which live outside the world's affluent societies (states). What Speth does, for very good reasons, is what most cosmopolitans focusing on climate change do: they use cosmopolitanism, which by definition is about the fundamental moral equality of all people *regardless of nationality*, but then they choose to ignore many people *because of their nationality*, despite those people's role in causing climate change or indeed their ability to help effect its mitigation.

Governments and policy-makers are largely ignoring the consumption habits of millions of affluent people in developing countries contributing to greenhouse gas emissions (see Chapter 6). Can we justify in ethical or practical terms what those affluent people (and affluent people in the developed world, of course) are doing? I expect that some affluent people in, say, China will say that what they are doing is not unethical, that China – and, by implication, *all Chinese people* – have no obligation to limit their activities that contribute to climate change, let alone being obligated to aid people in other countries who might suffer from it. However, one must challenge this national focus. As Singer (2003) shows:

> One of the clearest cases where [it] must be challenged is . . . climate change. Think about the difference that it makes to our conceptions of thinking ethically either within a community or globally once we understand that things that people do entirely within their own territory – like, for example, decisions about what kinds of vehicles we drive – could lead to making it impossible for, let's say, villages in Bangladesh to continue to farm low-lying delta lands where tens of millions of Bangladeshis make their living, because it may contribute to the rise in sea levels, which may mean that those lands become inundated and too salty to farm. Or it may contribute to changes in climate patterns in sub-Saharan Africa, which eliminates the reliable rainfall needed to grow crops.

Consequently, it should not be the case that we focus entirely on state obligations to cut greenhouse gases and to aid those suffering from climate change. We should focus more than we do now on the obligations of *people*, notably those who are affluent.

Even Singer makes the argument, true enough, about how the United States has used five times its collective per-capita share of greenhouse gases and China has used only three-quarters of its share. Singer's discourse lapses into that about states. Nevertheless, Singer's individual utilitarianism recognises that 'decisions and actions of human beings can prevent [extreme human] suffering' (Singer 1996: 26) and suggests that *all* the world's affluent have an obligation to act differently. Applying his principle – 'if it is in our power to prevent something bad from happening, without thereby sacrificing anything of comparable moral importance, we ought, morally, to do it' (Singer 1996: 28) – in the context of climate change seems to demand this, unless one assumes that frivolities and luxuries are more important than human survival and basic needs, not to mention ecological health.

Singer proposes two basic principles of fairness related to climate change: equal per-capita shares – it is hard to argue, although many have tried, for *un*equal shares – and the principle of 'you broke it, you fix it' (Singer 2003). He points out that this latter principle applies in the case of the affluent *states*. But it also applies in the case of affluent *persons*, including more than a few in the poor countries who have been polluting for generations. Those who are affluent ought to act, *regardless of whether the state in which they live is ethically or legally obliged to do so*. This, of course, means raising the sticky issue of ethical – and possibly new legal – obligations for affluent people in poor states. In the past we could overlook (from an environmental perspective) the relatively few affluent individuals in poor countries; their overall impact on the global environment was relatively low. That is no longer the case (see Chapter 6). We cannot continue to ignore them simply because interstate justice assigns no obligations to them. Thus Bradley Parks and Timmons Roberts (2006: 347) ask a fundamental question that needs to be explicitly addressed: 'Are states the relevant units of analysis in the study of climate justice?' As they importantly point out, 'the notion of the nation-state contributing to, being vulnerable to, and responding to climate change may obscure important intra-country distinctions. Many developing nations now have a sizable middle class that affects and is affected by warming of the Earth's atmosphere much differently than the rest of society' (Parks and Roberts 2006: 347).

Wolfgang Sachs (2002: 40) makes a cosmopolitan argument about climate change: 'The equal right of all world citizens to the atmospheric commons is . . . the cornerstone of any viable climate regime.' But even Sachs, one of the relatively few scholars attuned to the problem of the world's new consumers and the practical and moral recalculations they require, has a tendency to revert to state-centric discourse and solutions. He continues: 'a process allocating emission allowances based in per capita equal rights *to each country* has to be initiated' (Sachs 2002: 40, emphasis added). Similarly, Andrew Dobson (2005: 272), who sees the harms caused by climate change as inherently unjust, generally argues for a cosmopolitan response to the problem. But, like that of many other cosmopolitans (and of course all communitarians), his discourse is often about states. To wit: 'If global warming is principally caused by wealthy *countries* [emphasis added], and if global warming is at least a part cause of strange weather, then monies should be transferred as a matter of compensatory justice [to developing countries experiencing flooding] rather than as aid charity'. Even Vanderheiden (2008), who argues for cosmopolitan justice with due regard to both individual and collective responsibility, identifies the argument that 'the design of a global climate regime must allocate emissions caps to nations themselves and not to particular individuals, and assess liability at the national rather than individual level' (Vanderheiden 2008: 145). This moves back in the direction of states once again.

The proposal for actualising global justice described in Chapter 7 – a cosmopolitan corollary to international doctrine – while also retaining existing roles for states and movement towards action on their responsibilities as national communities, aims to go beyond these proposals. It supplements them with an approach that is based wholly on the rights and responsibilities of individuals – a cosmopolitan and global conception of climate justice – while calling on states (in part because there is no realistic choice to do otherwise) to act as the facilitators of global climate justice.

Jouni Paavola (2005) has gone some way towards bridging the gap between international and cosmopolitan (or cosmopolitan-like) responses to climate change. His focus is on adaptation to climate change, and he notes that international governance in this respect has centred on justice among states rather than justice for other actors. Paavola (2005: 319) argues that 'the greatest problem in the current institutional framework is its failure to address responsibility for climate change impacts [which] has to be addressed to create a functioning system for compensating for climate change impacts and assisting adaptation'. This is an important point. Indeed, it is the responsibility to which he refers that makes climate change a matter of justice. But one thing that is interesting about Paavola's discussion, and that makes it important to highlight here, is both what he has done and what he has not done. Unlike many other observers who focus purely on the injustices arising from climate change among states, Paavola is careful to note that *non-state actors within impacted states* suffer the effects of greenhouse gas emissions from *developed states*: 'the most significant issue of cross-level distributive justice [which considers non-state actors, such as households and communities] is the responsibility of developed countries for the impacts of their greenhouse gas emissions' (Paavola 2005: 317); 'responsibility for climate change impacts is primarily a cross-level distributive justice issue [so here he is taking what we might call a cosmopolitan approach] between developed countries [although here he takes a more statist perspective] and households, communities and organizations that are harmed by climate change impacts [again, back to what might be considered cosmopolitan], although it also involves distributive justice between the states' (Paavola 2005: 316) (this latter reference to purely state-centric conceptions of climate justice). And he says this even more explicitly: 'the climate change regime does not address the key issue of distributive justice – the responsibility of developed countries for the impacts of their greenhouse gas emissions [a purely state-centric statement]. These emissions cause climate change impacts that burden vulnerable people in developing countries who have not contributed to climate change and have little capacity to deal with it' (Paavola 2005: 310), a cosmopolitan statement, albeit focused on vulnerable people in vulnerable (poor) states.

If we take it all together, what Paavola seems to be saying is that developed states are responsible for their impacts on developing states *and* other actors. He also alludes to developed states' responsibility to other developed states and their citizens. But what seems to escape the equation altogether – this is common, but here it applies to someone who is uncommonly taking a more sophisticated, less state-centric approach – is the growing number of affluent people *outside* developed states who are now contributing greatly (indeed massively) to climate change, and his latter statement ignores the burdens experienced by people in developed states (for example, poor people in Britain not aided by their own government who might suffer the effects of greenhouse gas emissions by people or other actors in the United States, or, for that matter, emissions by wealthy people in Brazil). It forces one to ask: who is more responsible for the suffering of someone in, say, India: a lavishly well-off fellow Indian, or an ordinarily well-off person – or a not so well-off person, or even a badly off person – in, say, Britain? So I think we need to modify what Paavola says to read more like this: responsibility for climate change impacts is primarily a cross-level distributive justice issue among *all actors* causing climate change impacts and *all actors* harmed by climate change impacts. Naturally this requires qualification – not all actors are equally responsible because their circumstances and capabilities vary – but what it does is follow Paavola's movement away from a purely statist approach to climate justice without taking a step back by not fully comprehending cosmopolitan priorities: people matter and matter equally, *ceteris paribus*, and they have both responsibilities and rights, regardless of the states in which they live. Doing this would equitably implement Paavola's own prescription to bring into international environmental governance 'another justice dimension: justice in cross-level interactions between states and non-state actors such as individuals and their communities and organizations' (Paavola 2005: 312).

COSMOPOLITAN DISCOURSE

As will be argued in Chapter 7, the international discourse about climate change will have to change if the climate change regime is to be reformed and strengthened. At this point it may be worth spending a bit of time looking at that discourse and the assumptions that we make when thinking about climate justice. Let us start with this statement by Mark Smith (2006: 63), which captures many of the themes discussed so far: 'Most people in the industrialised world have, at least up to now, been able to afford to be complacent, but poor people in developing countries do not have that luxury and are having to adapt to the impacts of climate change.' What does a statement like that convey? Some people may read it and think, 'Clearly the industrialised world has got do more', and might think of the way that former American president George W. Bush spent eight years in office trying to protect the United States from requirements to reduce greenhouse gas emissions. That would be a rather statist interpretation, but it would be one that is very reasonable and highly sensitive to many of the international injustices inherent in extant responses to climate change.

Another way of interpreting Smith's statement would be less statist and more focused on people per se. From this perspective, we might get the message that people in rich countries are lucky to be able to ignore or cope with climate change relatively easily, whereas the poor people of developing countries are enduring substantial hardship to adapt to climate change. But if one reads the quotation very carefully it reveals, albeit without emphasising it, two groups of people who are routinely overlooked in discussions about climate change generally and climate justice in particular. It refers not only to 'most people in the industrialised world' but also implicitly to a relatively few *poor people of the developed world* who can *not* afford to be complacent (such as the poor people of New Orleans who to this day suffer the effects of Hurricane Katrina in 2005), and it points not only to 'poor people in developing countries' but, if we read it carefully, also to *rich people in developing countries*. If we spend some time thinking about the statement, and try hard not to focus too much on states, on the one hand, or only on the affluent people of the affluent states, on the other, we see the true sources of climate change: mostly the affluent people of the *world*, full stop. And we see those who suffer the most from it: poor people of the world, including some poor people in rich countries but still mostly poor people in poor countries where there is little adaptive capacity.

Smith's statement is a far cry from the more common statement, which is something like this: rich countries caused this problem, ought to clean it up and ought to pay the poor countries for their suffering. Most of us would agree with that and might even interpret Smith's statement as essentially saying the same thing, or certainly justifying this sort of conclusion (which it does). But the statement is also something short of really focusing on people, such as: rich people everywhere (regardless of the country where they live) have, at least up to now, been able to afford to be complacent, but poor people, especially in developing countries (but not only there), do not have that luxury and are having to adapt (or will do soon) to the impacts of climate change. That is a different kind of discourse. It is more decidedly cosmopolitan insofar as it draws our attention to the unjust pollution of rich *persons* and the unjust plight of poor *persons*. It suggests a different kind of world politics and a different kind of climate policy.

Communitarians will say that obligations obtain only within one's own political community – one's own nation or state. However, in the environmental area, and especially in the light of the causes and consequences of climate change, everyone is living in one interdependent community. As Singer (2004: 197) reminds us, 'when different nations led more separate lives, it was more understandable – though still quite wrong – for those in one country to think of themselves as owing no obligations, beyond that of non-interference, to people in another state. But those times are long gone. Today greenhouse gas emissions alter the climate under which everyone in the world lives.' Rights and responsibilities are associated with this reality. Everyone has a right not to be harmed by the pollution of others, whether they be next door or on the other side of the planet, at least if the polluters have the ability to control their pollution. Everyone, and especially those of us who are most capable (usually the most affluent), also has an obligation to act in ways that do not violate those rights. That we are living in this single world also suggests that we have obligations to aid others, even those very far away, whom we have harmed or will harm.

CONCLUSION

Climate change cries out for a cosmopolitan response. It is a global problem with global causes and consequences. The idea that states can continue to control what happens within their borders is much less meaningful because 'some of the most fundamental forces and processes that determine the nature of life chances within and across political communities are now beyond the reach of individual nation-states' (Held 2000: 399, quoted in Vanderheiden 2008: 89). From this perspective, climate change is one contemporary phenomenon that creates 'overlapping communities of fate' requiring new cosmopolitan institutions (Vanderheiden 2008: 89). Appropriately, the notion that individuals have rights to some environmental minimum, and not to suffer from environmental harm caused by others, has already found its way into the climate change regime. However, at present those who have a recognised duty to protect or at least to limit violations to those rights are states (and, sometimes, corporations). Generally speaking, the most that individuals must do at present is pay taxes and comply with minimal regulations imposed by some developed-country governments. Where cosmopolitan justice is especially important is in locating obligation – to stop harming the environment on which others depend and to take steps to aid those who suffer from harm to the environment – on the shoulders not only of governments but also of individuals. As Attfield (2003: 182) points out, 'the global nature of many environmental problems calls for a global, cosmopolitan ethic, and for its recognition on the part of agents who thereby accept the role of *global citizens* and membership of an embryonic global community'. Cosmopolitan justice, and the associated obligations, should therefore at minimum supplement the traditional international justice view and its associated obligations – although it should not dilute the common but differentiated responsibilities of states.

One curious attribute of the expansion of environmental justice to international affairs, and to climate change in particular, is that those who were initially the focus of environmental justice – people and communities *in the developed countries*, the United States in particular, suffering from poverty and discrimination – seem to have been cast aside. When the discourse about environmental justice shifted from people and local communities to states and international relations, so too did the focus of scholars and policy-makers. The question shifted to a great extent from 'what do we owe poor people suffering from pollution of rich people?' to 'what do we owe poor *states* suffering from pollution of rich *states*?' The latter question is perfectly reasonable; international justice demands that we ask this question and act on answers to it. But to an increasing extent that question diverts our attention – and the policy actions by international organisations and the aid money and so forth – away from people to states. Indeed, by focusing on the former question we would deal with what the latter one presumably intends to deal with: most people within poor countries suffering the effects of pollution. But the latter question can very easily overlook them. Consequently, it would not be unreasonable to conclude that the apparently important and commendable extension of environmental justice to international relations should be lauded while also concluding that we ought to pause to ask where this is leading us. This is not to abandon international environmental justice as an important theme of international relations and an important concept to implement, but instead to suggest that we need to be careful that we do not forget about people on the ground who are, or who ought to be, the ultimate concern of environmental justice. It is the cosmopolitans who routinely remember these people. As such, a cosmopolitan perspective can suggest important new directions for the climate change regime.

As Patrick Hayden (2005: 3) says, cosmopolitanism is simultaneously an ethical project and a political project: 'As an ethical project it seeks to establish the extent and content of, and justification for, moral obligations concerning the well-being of every individual person. As a political project it is intimately connected with debates about the appropriate form of political community, schemes for legal institutions and procedures, and practices of humanitarian assistance on a global scale.' Given the impact of climate change and the growing number of relatively affluent persons causing it, there is a very strong case for institutionalising cosmopolitan ethics in climate diplomacy and policies. This can be fully justified *ethically*, but the most important point for everyone (even communitarians and statists) may be that cosmopolitan considerations are *practically* and *politically* necessary to foster effective actions to combat climate change, as the next two chapters argue. In Chapter 7 I propose a cosmopolitan corollary to the doctrine of international climate justice that attempts to encompass these ethical, practical and political considerations. But first, in Chapter 6, I look more deeply into the role played by persons, including the many of them not given much consideration even in most cosmopolitan assessments of, and prescriptions for improving, the climate change regime: the world's new consumers.

NOTES

1. While Rawls (1971, 1999) does not say so, several scholars, notably Brian Barry, Charles Beitz and Thomas Pogge, have argued that Rawls's premises lead to the conclusion that resources and wealth ought to be redistributed to the world's least well-off people (cf. Jones 1999: 2).
2. I invoke cosmopolitan or cosmopolitan-like arguments, but I cannot say whether all of these scholars would call themselves cosmopolitans.
3. Attfield (1999: 205) advocates a consequentialist variant of cosmopolitanism based on needs.
4. Caney (2005a) points out that knowledge of climate change has been sufficient since roughly 1990 to neutralize arguments of ignorance about its causes and severity (see Chapter 6).

CHAPTER 6

AFFLUENCE, CONSUMPTION AND ATMOSPHERIC POLLUTION

One reason why climate change is a matter of *global* justice – of *cosmopolitan* justice, which fully considers the rights and duties of individuals – is that millions of people geographically and temporally distant from the sources of the global warming suffer from its consequences. Another reason is that the persons who are presently causing future global warming no longer live almost exclusively within the states that have historically caused it. Until quite recently we could talk, in both moral and practical terms, about climate change as a problem caused by the world's developed countries and their citizens. They were by far the primary sources of greenhouse gas pollution and thus (if we focus solely on causality) the logical bearers of responsibility to end that pollution, to make amends for it and to aid those who will suffer from it. The climate change regime, insofar as it acknowledged this responsibility, is premised on this notion of developed states – and, indirectly, their people and commercial entities – having primary responsibility for addressing this problem. However, climate change is no longer solely or even predominantly caused by the relatively affluent people of the world living in the affluent states of the world. Increasingly pollution of the atmosphere comes from the growing number of people in developing countries who are joining the long-polluting classes of the developed states. The developing countries together now produce fully half of the world's greenhouse gases.

Given the developing countries' large populations, this change does not alter the *international* moral calculus very much. After all, their national per-capita emissions remain well below those of the developed states. What is different is the growing number of 'new consumers' (cf. Myers and Kent 2004) in the developing world, many of them extremely affluent persons, who are living lifestyles similar to those of most people in the developed countries. Indeed many of these new consumers are living more like the wealthiest classes in the industrialised world. These new consumers now number in the hundreds of millions. They are producing greenhouse gases through voluntary consumption at a pace and scale never experienced in human history. While societies in the West are starting to make changes to limit their greenhouse gas emissions, the new consumers in developing countries are going in the opposite direction. The consequences for the atmospheric commons, and indeed for other environmental commons, are already shaping up to be monumentally adverse.

Most literature on climate justice overlooks the rapidly increasing share of the cause of climate change coming from this growing number of new consumers in developing countries. These people have been omitted from climate change agreements, diplomacy and most discourse about climate justice. At present, they face few if any legal obligations to mitigate the harm they do to the global environment, and they have so far almost totally escaped moral scrutiny. Insofar as the world's new consumers have been scrutinised, it is by a relatively few cosmopolitan scholars (e.g. Caney 2005a, 2006b, 2008). If solutions to climate change are to be found, this will have to change, not least because 'old consumers' in the developed states are watching these new consumers consume in ways that the old consumers are being told they must stop for the sake of the Earth's atmosphere. This makes it politically nigh impossible for the old consumers' governments to force new greenhouse gas cuts upon them. As long as the new consumers are able to hide behind their states' overall poverty, practical and politically viable solutions to climate change will remain elusive.

To be sure, the need for cosmopolitan justice and the responsibility of affluent individuals obtains regardless of how many affluent people live in developing countries; affluent people in developed countries should be reducing their pollution regardless of what is happening in the developing world. However, the rise of the new consumers makes cosmopolitan, global justice more vital than ever. The emergence of a significant number of affluent polluters in developing countries is a practical reason for stressing cosmopolitan justice, not the ethical rationale. What most Americans and most Britons should have been doing all along, and ought to be doing right now, is also an obligation of growing minorities in Argentina and Thailand. And the rise of the new consumers offers an opportunity for the governments of developing states in effect to join in greenhouse gas limitations without having to take on new ethical or legal burdens. They can do this by demanding that affluent polluters within their borders behave as affluent polluters everywhere should behave.

The rise of hundreds of millions of new consumers in the developing world may be the most important environmental phenomenon in human history. This chapter identifies and examines their role, in so doing highlighting the ethical and practical advantages of viewing climate change through a cosmopolitan lens. The first section describes the growth of the new consumer class. The second section explores some of the normative implications of this phenomenon, particularly the significance of spreading affluence for how we think about justice and climate change. The aim here is not to mediate among philosophical viewpoints, but simply to show that there is ample ethical justification for saying that, in the context of climate change, obligations of justice lie with capable and responsible persons *everywhere*, not just with the most capable states and their citizens.

THE NEW CONSUMERS

Until recent decades it was possible to talk in terms of developed and developing countries, North and South. However, in the context of climate change this is no longer an accurate terminology. As Wolfgang Sachs and Tilman Santarius (2007: 67) point out, to talk in terms of North and South hides the reality that 'the social cleavage within countries may be at least as dramatic as the gap between countries . . . A dividing line runs between the globalised rich and the localised poor . . .'. Very importantly, for an accurate, current understanding of climate change, we need to recognise that hundreds of millions of affluent (that is, middle-class and upper-class) people now live in developing countries, and that they have the power to consume as much as, and sometimes more than, people in developed countries (Sachs and Santarius 2007: 67–8). This huge number of increasingly affluent people in developing countries has been overlooked in part because it is a recent phenomenon.

To understand the importance of the world's new consumers for climate change, it is helpful to get a picture of their numbers. Although different studies arrive at different counts of new consumers, together they give us a picture of growing pollution. According to one study, if we set the level of relative affluence in developing countries quite low, at about the level where people can meet their basic needs and start to consume in ways akin to those in developed countries (that is, above about $2,500 per year) – roughly at a level where ethical responsibility for emissions begins – we arrive at over one billion new consumers worldwide (Sachs and Santarius 2007: 78). According to another study, if we set the amount at $7,000 per year (based on purchasing power parity), akin at least to the purchasing power of the lower-middle classes of Western Europe, the number of new consumers is over 800 million, compared to about 900 million established consumers in the developed countries (Sachs and Santarius 2007: 78). China and India together account for over one-fifth of the 'global consumer class', a number that is approaching 400 million and that exceeds the number of people living in Western Europe (Sachs and Santarius 2007: 78). The latter, on average, still consume more per capita, although to say this ignores the large number of very affluent people in China and India who consume *more* than most Western Europeans. As Sachs and Santarius (2007: 78) summarise these trends, generally speaking, this 'transnational consumer class resides half in the South and half in the North. It comprises social groups which, despite their different skin colour, are less and less country-specific and tend to resemble one another more and more in their behaviour and lifestyle.'

According to an analysis by Norman Myers and Jennifer Kent (2003: 4963), the environmental impacts of increasing consumption in the world are rising very quickly because 850 million consumers in developed countries were joined quite recently by nearly 1.1 billion new consumers in a number of developing and transition countries.[1] Myers and Kent (2003: 4963) define the new consumers as a four-member household with purchasing power of $10,000 per year. This level of purchasing power represents 'a degree of affluence that enables wide-ranging purchases such as household appliances and televisions, air conditioners, personal computers, and other consumer electronics, among other perceived perquisites of an affluent lifestyle' (Myers and Kent 2003: 4963). Myers and Kent point out that of even greater environmental significance is the fact that many of the new consumers are buying cars and shifting to diets heavy in meat, both actions that contribute greatly to greenhouse gas pollution. In the twenty countries that Myers and Kent analysed, the new consumers own nearly all the cars, and the numbers of cars purchased were increasing in leaps and bounds, explaining why cars are expected to comprise the fastest-growing energy-use sector in the next two decades or so (Myers and Kent 2003: 4965). The national incomes of new consumers are 'far greater than national averages' and these people 'have a far-reaching impact on economic activities nationwide, and hence on environmental repercussions' (Myers and Kent 2003: 4963), showing the practical limitations of thinking in terms of *national* per-capita incomes, as is routine in discussions of climate change generally and climate justice particularly.

NEW CONSUMERS IN CHINA AND INDIA

China and India exemplify the rise of the new consumers. In the case of India, new consumers make up only one-eighth of the population but, incredibly, possess two-fifths of the country's purchasing power, and they are responsible for 85 per cent of private car purchases (Myers and Kent 2003: 4963). The number of new consumers in India was estimated by Myers and Kent (2003: 4966) to be about 132 million in 2000 (13 per cent of the population), with that number expected to grow to about 205 million (18 per cent of the population), but possibly 225 million, by 2010. In India – where headlines read 'Splurge. Because you can now' (Karunakaran 2002) – the new consumers' energy consumption causes carbon-dioxide emissions that are fifteen times those of the remainder of the country's population (Myers and Kent 2003: 4963). Not surprisingly, then, despite most of its people living in poverty and consequently emitting very little greenhouse gas pollution per capita, India is likely soon to become the third largest national emitter of carbon dioxide, overtaking Russia (Global Carbon Project 2008).

Des Gasper's description of modern India helps to demonstrate that the problem of affluence is not only that of indifference and pollution among affluent people living in developed countries:

> If one walks the streets of a metropolis in India nowadays, one can sometimes get a feeling that not only the rich but also increasing numbers of the professional classes have morally seceded from the nation. Many seem to live the same in various ways as Indian professional emigrants abroad, or foreign tourists, or those same tourists when back home in the North. The smartly dressed well-to-do proceed from gleaming cool office or home interiors, communicating to each other on their cell phones, through streets with many wretched begging people whom they generally ignore, to shops and hotels full of luxuries and imports from America, Britain and Singapore for which they can evidently afford to pay world prices. In the 1990s, while consumerism reached new levels in India, public sector expenditures were squeezed. The affluent seem to have become semi-detached in their own country, inhabitants of a quasi-apartheid system moving further in the direction of Brazil or South Africa. In effect they declare that if the elites and middle classes of other parts of the globe are entitled to live in a certain way, then so are they . . . (Gasper 2005: 6–7)

In South Asia more generally, the now-very-large number of upper-middle-class people use as much energy as do people in that class in developed countries, creating an unevenness between the new consumers and the majority of people within South Asian countries that mirrors disparities between people in the developed countries and people in the developing countries generally (Sachs and Santarius 2007: 77). To be sure, these newly affluent people are not yet collectively causing the same amount of harm as the rich countries' affluent classes, but this does not absolve them of obligation. And, practically speaking, the trend is towards the new consumers fully joining, and possibly replacing, old ones as the main drivers of climate change.

Developments in China further show how the international (and communitarian) perspective fails to account for ongoing developments. In 2006 China overtook the United States to become the largest national source of greenhouse gas emissions (Netherlands Environmental Assessment Agency 2008), and its emissions will continue to increase (Global Carbon Project 2008). The Chinese Academy of Sciences predicts that, without dramatic changes in business as usual, by 2030 China's greenhouse emissions will more than double, rising from about 5.1 billion tonnes of carbon in 2005 to as much as 14.7 billion tonnes (compared to 31.2 tonnes for the entire world in 2007) (Brahic and Reuters 2008). To be sure, much of the pollution from China is a consequence of feeding the material appetites of Western consumers. But China is also the country where the most new consumers reside, with hundreds of millions of Chinese joining the global consumer culture. Indeed, China has 'the greatest scope for generating more new consumers in the future' than any other country, with their number predicted to increase from just over 300 million (24 per cent of the population) in 2000 to about 600 million (44 per cent of the population) in 2010 (Myers and Kent 2003: 4966). The number of passenger cars in China doubled every thirty months during the 1990s, and some reports anticipate that the number of cars in China could increase from 1.1 million in 1990 to about 56 million by 2010 (Myers and Kent 2003; Gallagher 2006: 4966). Official Chinese estimates predict that the total number will reach 140 million by 2020 (*China Daily* 2004), which is roughly the same number of cars as in the United States today. Goldman Sachs predicts that within two decades there will be 200 million passenger cars in China – more than in the United States (Dyer 2006). China's poor will not be driving these cars.

According to one report, there are an estimated 450 million people in China's eastern provinces with individual purchasing power exceeding $7,000 annually ($6,000 is considered a threshold after which the buying of cars 'begins to take off') (Economist 2005). Jonathan Garner (2006: 73) predicts that the number of Chinese households earning more than $10,000 per year will increase from 3.8 million in 2003 to 151 million in 2013 (compared to a US figure of 102 million in 2003). Between 2004 and 2013, the number of urban households in China 'able to make discretionary consumer purchases beyond meeting basic needs' could increase to 212 million from 31 million, rising from 17.4 per cent of households to 90.6 per cent (Garner 2006: 73).[2] Many Chinese are consuming and living more like the stereotypical American conspicuous consumer, and 'the locomotive of the global economy in terms of incremental annual consumption demand [is changing] from the US consumer to the Chinese consumer' (Garner 2006: 13). Chinese tourists are now common in the world's shopping capitals, where they snap up luxuries, such as very expensive jewellery, high-status fashions and the latest consumer electronics. All of this points to the extent to which China's economic rise will have a 'marked impact on environments both national and global, and a good part of these impacts will reflect the rise of the new consumers' (Myers and Kent 2003: 4966). This is a key point, because the Chinese government is always anxious to focus on the many remaining poor people in China and on the impact of exports to the West as contributors to China's total emissions, presumably expecting that everyone will ignore the millions of affluent Chinese behaving just as their counterparts in the West have done for decades.

The total population of new consumers in China and India together could reach 825 million by 2010 (Myers and Kent 2003: 4966). However, these two countries accounted for only 41 per cent of the new consumers (in 2000) counted by Myers and Kent (2003: 4967), leaving tens of millions more (for example, 74 million in Brazil, 63 million in Indonesia, 68 million in Mexico) that must be added to this total number, along with their collective contribution to climate change. While the number of new consumers adds up to more than the total number of 'old' consumers in developed countries, their total collective purchasing power – and thus their potential polluting power – is still below that of the old consumers. Nevertheless, Myers and Kent (2003: 4967) argue that there is a very 'sizeable "North" in the "South"', with the new consumers constituting 'a major consumer force in the global economy, just as they are becoming a front-rank factor for the global environment. China's environmental impact [alone] could eventually match that of the US' (if it is not doing so already). And the *rate* of consumption among new consumers is rising, meaning that they are moving closer to people in the West just as those people are being asked to reduce their own rates of consumption.

REVEALING PEOPLE'S CONSUMPTION

As we have seen, the developing countries in aggregate have overtaken the developed countries as the largest national sources of greenhouse gases. But what is most important for global justice is that soon the number of people in developing countries who are consuming in the same way as people in the developed world will exceed the number of consumers in the industrialised countries (if it has not already done so). Myers and Kent (2003: 4967) believe that the world's new consumers are unlikely to change their consumption habits before the old consumers of the affluent countries do so, and they believe that it is the latter who ought to do so first. Hence the perennial political problem of states obtains with people, too: who goes first? A cosmopolitan viewpoint might suggest that all the world's affluent consumers ought to go at the same time. To be sure, historical responsibilities associated with climate change are new for these new consumers, but it is a major flaw of the climate change regime that these (and all) *people* are not given more prominence, and indeed that they are largely ignored in official discourse and international legal instruments. The point that Myers and Kent highlight is the importance for global justice of dissecting arguments about climate justice to look at what different people, or perhaps more easily different groups and classes of people, are contributing to climate change.

Per-capita emissions ought to be looked at in their own right, not just in terms of states. That is, it is not enough (and probably wrong) to think, speak and act in terms of per-capita emissions in (for example) Australia versus Indonesia. Per-capita emissions in the former far exceed those in the latter. From this it is concluded that Australians are bad and Indonesians are not, or at least that the former are 'more bad' (or less good) than the latter. But to think of per-capita emissions from this state-centric perspective creates an ethical problem and an increasingly practical one. It means that people living in states with low *average* per-capita emissions of greenhouse gases are not part of the moral calculus undergirding climate justice, nor are they part of the practical calculations about where emissions arise. Per-capita emissions, as normally conceived, create a screen behind which many millions of affluent people can hide, both morally and practically. It is important to bear in mind that, while average per-capita emissions in, say, Austria probably give a generally good reflection of how nearly all Austrians actually consume and pollute, this it not the case in many developing countries. The top *hundred million* consumers and polluters in China, for example, have lives that are utterly unrecognisable when compared to the hundreds of millions of people in China who live in what, by any Western measure, is abject poverty. Thus, for us to invoke average per-capita comparisons while ignoring these realities prevents us from seeing climate change for what it really is: a problem unnecessarily and unjustly caused by certain kinds of people (namely, the affluent and capable) nearly *everywhere* – at least everywhere that affluent people live.

Sachs is one of the few people who have been vocal about obligations of *all* the world's affluent people. He is critical of our usual focus on what he calls the 'zombie category of the nation-state':

> The nation-state is an artefact. It's a category that does not reflect reality adequately, but we are stuck with it for diplomatic reasons . . . Most importantly, what is being covered up by that artefact is that the real gulf in the world is not between the Northern and the Southern countries, but between the global middle class and the marginalized majorities, and that a quarter to a third of the global middle class is sitting in the South . . . You have a Germany sitting right in India. Germany has 82 million inhabitants, not all of them are really rich; I mean, there are easily 70 million middle class in India. (Sachs 2001)

In Sachs's view, the real equity (social justice) issue is not the one among states: 'The real equity issue is between the global middle class and the marginalized majority. They are affected by the climate by being the victims of climate change. Now that is the serious equity question. It's a different level. There are two levels of equity in the climate discussion. And that's the more serious one' (Sachs 2001).

Sachs (2002: 19–20) shows how the concepts of 'North' and 'South' ('developed' and 'developing' countries) no longer reflect political and economic realities. Given that the 'South' includes both wealthy Singapore and poor Mali, it is hard to see what the countries of the South have in common (apart from defending and reinforcing state sovereignty). More important is the way the ideas of North–South and rich–poor states obscures

> the dividing line in today's world [that] is not primarily running between Northern and Southern societies, but right across all of these societies. The major rift appears to be between the globalised rich and the localised poor. The North-South divide, instead of separating nations, cuts through each society . . . It separates the global consumer class on the one side, from the social majority outside the global circuits, on the other. This global middle class is made up of the majority of citizens in the North, along with a varying number of elites in the South . . . In attracting resources, their geographical reach is both global and national . . . Particularly in Southern countries, market demand for resource-intensive goods and services stems mainly from that often relatively small part of the population, which commands purchasing power and is therefore capable of imitating the consumption patterns of the North. As a consequence, the more affluent groups in countries such as Brazil, Mexico, India, China, or Russia use about as much energy and materials as their counterparts in the industrialized world . . . Reduction of the ecological footprint of the consumer classes around the world is not just a matter of ecology, but also a matter of equity . . . since the excessive use of environmental

space withdraws resources from the social majority in the world [and] wealth on
one side is at times co-responsible for poverty on the other. (Sachs 2002: 20)

The upshot is that 'fairness demands reducing the ecological footprints
of the consumer classes in North *and* South' (Sachs 2002: 23, emphasis
added).

The real locus of climate action, and of climate justice, is at the
individual level. As Steve Vanderheiden (2008: 165) points out, despite
the illusion that individual actions causing greenhouse gas emissions
'have no effect and therefore appear to be entirely faultless, we must
recognize that this false appearance is due to a mistake in disaggregating
individual contributions from group-based harm. [Individuals] may be
held responsible for problems that do not appear to be caused by any
person's individual action but yet *collectively* cause harm.' However,
this is something that is remarked upon relatively rarely in the literature
on climate ethics and policy. One might retort that we have always
downplayed the role of some individuals; many rich elites in poor
countries have always avoided responsibility. For example, corrupt
officials and their extended families in far too many poor states have
for decades siphoned off development aid to increase their personal
fortunes. However, that kind of behaviour has always been viewed as
wrong and usually illegal, sometimes even in the states where it occurs.
In contrast, for the growing number of economic elites in the developing
world to pollute the global environment, which will contribute to
widespread human suffering, including among their compatriots, is
rarely subjected to comment, especially compared to the frequent
comments about ordinary people in wealthy countries causing climate
change. Indeed, nowadays the highly consumptive and polluting behaviour of new consumers in the developing world is being strongly
encouraged by global capitalists seeking to benefit from increasing
buying power in the developing world. Such behaviour is seen as a
cure for global economic recession and the saviour of the world
economy.

As Lorraine Elliott (2006: 348) highlights, affluence is 'the primary
and disproportionate cause of global environmental degradation'. At
the turn of the century, the world's richest 20 per cent of people, the
majority of them living in the developed countries, were responsible for
86 per cent of all private consumption and 58 per cent of all energy
consumed (Elliott 2006: 348). Her point about affluence is very important: it is not the number of people that matters nearly as much for
the global environment as the number of affluent ones. But, given what
Myers and Kent (2004) say about the recent, very rapid increase in the

number of new consumers in the developing world, the notion that affluent people in industrialised countries are mainly to blame is becoming less true in practical terms. This reminds us, as robust cosmopolitanism does, to think less in terms of countries where affluence predominates and more in terms of affluent persons themselves, as well as the connections between their affluence and those people who suffer from it, people who admittedly live disproportionately (but not exclusively) in developing countries.

Thus, for both practical and ethical reasons, it is important to note the distinctions and relations among affluent people in poor states and affluent people in affluent states, the poor in poor states and the poor in affluent states, the affluent in poor states relative to the poor in affluent states, and so forth. Things start to look more complicated when we make these distinctions, but they also better reflect reality. The affluent (even the very wealthy) in poor states are ignored by the climate change regime, whereas the poor in a number of rich states are obliged, via their states' obligations under the climate change agreements, to bear the burdens of acting on climate change – even if they have contributed far less to it than affluent people in poor states who have no similar obligations, and even if they do not enjoy many of the fruits of the greenhouse gas pollution of their compatriots in the past – and even if they are harmed by those 'fruits'. This can be exemplified by the plight of a very poor person in the United States having no choice but to use a car to get to work, even when she would prefer to live near her workplace or travel by public transport, because the living and working infrastructure built during previous centuries leaves her no other choice but to drive a car – an old and highly polluting one at that. The aim of international climate justice today – of the climate change regime – is not directed at persons like this, at least not when we fully consider the causes of global warming. And the climate change regime misses nearly all the new consumers, a practical and normative problem that will increase year by year if not addressed.

AFFLUENCE AND CONSUMPTION THROUGH A COSMOPOLITAN LENS

One thing that seems unassailable from the perspective of world ethics and global justice is that greenhouse gas emissions required for subsistence take priority over other kinds of emissions, and ought not be subject to any kind of limit (cf. Shue 1993). In Dower's words (2007: 210), the 'right to subsistence surely takes precedence over the right to maintain affluency at the highest level'. This means that the 'luxury' emissions of rich people, regardless of where they live, ought to be a focus of greenhouse gas cuts. As James Garvey argues, if all the emissions in (for example) Rwanda are subsistence emissions and the bulk of emissions by people in the US Virgin Islands are luxury emissions, 'it's clear who has room for reduction and who doesn't. Arguing the point is as good as saying that some Rwandans should die so that some Virgin Islanders can recharge their mobile phones' (Garvey 2008: 81). It therefore seems self-evident that it is wrong for affluent people to harm unnecessarily the planet and the poor who are most dependent on it. Justice at least demands that we end the harm that we cause.

As Henry Shue has argued, while some will say that there is no obligation for me to help strangers whom I have not harmed,

> it is a very different matter if I have in fact wronged the person whose plight is under consideration – if that person's plight was caused by harm that I did. The question, ought I now to help someone whose need for this help results from harm that I myself inflicted? is radically different from the question, ought I to help a stranger whom I have never harmed? And the reason that the situation is so different when harm has been done is that one of the most basic principles of equity in every culture . . . is: *Do no harm*. One may or may not be expected to help in this or that context, but one is always expected not to harm (but for exceptional overriding circumstances). Consequently, the obligation to restore those whom one has harmed is acknowledged even by those who reject any general obligation to help strangers. Whatever one's obligation to help people with whom one has no previous connection, one virtually always ought to 'make whole', insofar as possible, anyone whom one has harmed. And this is because one ought even more fundamentally to *do no harm in the first place*. (Shue 1995: 386, emphasis added)

This suggests that one basis for our obligations as affluent people to act is quite fundamental, and to argue otherwise would contradict ethical norms nearly everywhere.

Various species of cosmopolitanism point to more reasons why climate change demands different behaviours, and ultimately different policies.

VITAL INTERESTS, COMMON HUMANITY AND RELATIONSHIPS OF OBLIGATION

Two principles of justice described by Brian Barry (1998: 148–9), based on the premise that what is just is what the least well-off could not reasonably reject, are germane to climate change: (1) personal responsibility and compensation, and (2) the priority of vital interests. According to the first principle, people may fare differently 'if the difference arises from a voluntary choice on their part; conversely, victims of misfortunes that they could not have prevented have a prima facie valid claim for compensation or redress', and 'where the voluntary act of some person (or persons) is the cause, redress should be looked for in the first instance from that source' (Barry 1998: 148). According to the second principle, 'the vital interests of each person should be protected in preference to the nonvital interests of anyone' (Barry 1998: 148). As we have seen, the first principle suggests obligations on the part of the world's affluent people because climate changes they help create cause harm to others (at least in the future). The second principle is especially provocative, requiring that the material luxuries of the rich be curtailed to limit harm and to provide resources for redistribution to help protect the vital interests of the poor.

Similarly, in a discussion about the importance of international climate justice, Shue makes a claim that is just as well suited to cosmopolitan climate justice: 'it is unfair to demand that [the poorest] be sacrificed in order to avoid our sacrificing interests that are not only not vital but trivial' (Shue 1992: 394). This is the heart of the matter to a great extent: after a point that meets our needs and then a bit, the world's affluent are contributing to climate change for relatively trivial reasons at the expense of the vital interests of the world's poor. As Caney (2005a: 769) puts it, the most advantaged individuals 'can perform the roles attributed to them, and, moreover, it is reasonable to ask them (rather than the needy) to bear this burden since they can bear such burdens more easily'. Caney (2005a, 2005b) prioritises the interests of the global poor, arguing that 'the least advantaged have a right to emit higher GHG [greenhouse gas] emissions than do the more advantaged in the world' (Caney 2005a: 770). The upshot is that justice does not permit poor *persons* to be told to sell *their* blankets in order that rich *persons* may keep *their* jewellery (cf. Shue 1992: 397).[3]

Individual obligation and action can be justified from other perspectives. For example, Onora O'Neill's (1988) Kantian cosmopolitanism locates duty in individuals, who share a common humanity, suggesting that at least we ought not to undermine the capacity of others to be independent moral agents. Bearing in mind that climate change will affect human rights, particularly the most basic rights to sustenance and even survival, another way of looking at climate justice is from a human-rights perspective (see Pogge 2002b). One way that climate change violates human rights, as Thomas Pogge (2002c: 184) suggests, is by denying individuals a choice about whether to live in a degraded environment. Many people, perhaps all of us, will be forced to do so. As Ciaran Cronin and Pablo De Greiff (2002: 18) describe this perspective, a person's human rights are not only moral claims *on* any institutional order imposed upon that person, but also moral claims *against* those – especially, the more influential and privileged – who collaborate in its imposition. Since human rights-based responsibilities arise from collaboration in the coercive imposition of any institutional order in which some persons avoidably lack secure access to the objects of their human rights, it follows that there are transnational obligations that fall primarily on the more influential and privileged agents (individual and collective) who collaborate in the imposition of the current international order [which here is the 'order' that results in climate change] since it satisfies this condition.

Note the way that concern for human rights can place obligations on the shoulders of privileged individuals.

Jon Mandle (2006) advocates a very minimalist form of cosmopolitanism based on protecting basic human rights while otherwise condoning the role of the state. He argues that 'beyond the protection of basic human rights, global justice does not require higher levels of protection . . . and should not be subject to international enforcement' (Mandle 2006: 138). But even this minimalist formulation seems to support the protection – and enforcement of that protection – of basic rights associated with the atmosphere. We might think of this as a sort of corollary of the argument that people have rights to a stable and clean environment, or the right to sustainable development. David Schlosberg (2007: 94) points out that 'capabilities and functioning' are central here, because climate change undermines the ability of people to sustain themselves and undermines the integrity of their communities.

For Andrew Dobson (2003: 22), the utility of cosmopolitanism derives not from thinking in terms of humanity per se, but in 'specific communities of obligation – or "obligation spaces" – produced by acts of "globalization" (i.e. local acts with global consequences)', and he argues that 'we should recognize that these are communities of injustice first, and . . . that the remedy is therefore more justice as well as more democracy . . .'. Put another way, 'globalization is best regarded as a producer of this political space of asymmetrical obligation' (Dobson 2003: 30), and, insofar as justice is about asymmetries in shared political spaces, processes of globalisation extend the reach of justice to the entire world. Dobson points out that this is especially true in the context of environmental change. He quotes Judith Lichtenberg on this point: 'the relationships in virtue of which the earth now constitutes one world are . . . pervasive and far-reaching[, and] actions . . . may have harmful consequences without any direct involvement between agents and those affected' (Lichtenberg 1981: 87, quoted in Dobson 2003: 30). This leads to obligations on the part of those who pollute. Dobson (2003: 31) uses climate change as an illustration: 'If global warming is at least a part cause of strange weather, then monies should be transferred as a matter of compensatory justice rather than as aid or charity.' From this perspective, then, climate change 'changes the source and nature of obligation' by creating what Dobson calls a 'post-cosmopolitan relationship of actual harm' (Dobson 2003: 31).

Recall Dobson's argument (2006) about 'thick cosmopolitanism' in the preceding chapter. Injustice is manifest when we cause harm to others. The connection between cause and resulting harm is that which creates injustice and motivates action (or should do). If, as Dobson (2006: 174) suggests in an example, floods in Mozambique are 'in part a result of the lifestyle of (some) members of . . . affluent societies, then . . . the only adequate response is the recognition of causal responsibility and the doing of justice'. Climate change and other environmental changes 'thicken the ties that bind us to "strangers", to bring these strangers "nearer" without having to rely on empathetically constructing them as surrogate neighbours . . .'. (Dobson 2006: 175–6). Everyone has an ecological footprint, and 'the relationships that go to make up an individual's ecological footprint amount to links in the chain of "causal responsibility" [which] is what makes "nearness" a more palpable reality than can be produced through reflecting on our common humanity' (Dobson 2006: 177). Dobson (2004: 11–12) argues that the '*nature* of the obligation is to reduce the occupation of ecological space, where appropriate, and the *source* of this obligation lies in remedying the potential and actual injustice of appropriating an unjust share of such space . . . Obligations are owed by those in ecological space debt, and these obligations are the corollary of a putative environmental right to an equal share of ecological space for everyone.'

Dobson asks us to look into our supermarket trolleys to see how wide our ecological footprint spreads, but in so doing he is referring to the footprints of the 'affluent societies', mentioned above, and of 'richer countries' (Dobson 2006: 177) and presumably people living there. What also needs to be said is that the trolleys in supermarkets of Hong Kong and Mumbai reveal similar footprints. With this said, it is worth noting that Dobson (2006: 177) regards the ecological footprint as 'a space of obligation' that demands justice. But Dobson's arguments require some qualification. For example, he says that 'anyone who consumes goods from Wal-Mart, Toys "R" Us, Tommy Hilfiger and Tesco [companies identified by Oxfam as driving down wages of workers, among other harms they inflict] is subject to the charge of complicity in transnational harm – the avoidance of which is a key cosmopolitan injunction' (Dobson: 2006: 177–8). It is this causal responsibility that will 'get us to do cosmopolitanism as well as believe it', because it binds us to people living far away (Dobson 2006: 178). But what of the many very poor people in the United States who shop at Wal-Mart because they cannot afford to shop anywhere else, and the poor who do likewise for the same reasons at Tesco in Britain? While those Western poor are causing harm along with affluent people in Beijing who are shopping at the same companies' supermarkets, responsibility of the Western poor is not necessarily the same – and surely it is not the same as that of people who shop at the more upscale Tommy Hilfiger shops in both rich and poor countries. It is very important to acknowledge that the ecological footprint of affluent people does not depend on where they put their feet (that is, their nationality), yet it is there that the debate about climate justice has been revolving, and from which it has been evolving, for decades.

POGGE'S COSMOPOLITANISM AND DUTIES NOT TO HARM

An extended discussion of some of Pogge's work is enlightening at this point because it helps to illustrate what is important about cosmopolitanism, and cosmopolitan philosophy, for climate change policy, and because it shows us how cosmopolitanism is quite often deployed in arguments about global justice and climate change. It is now routine for philosophers and even statespersons to argue that affluent people in affluent countries have obligations to the poor and destitute of the world. (Of course, communitarians might argue that no such individual obligation exists.) As an illustration, we can look at this summary of Pogge's (2002c) thesis by Gillian Brock and Darrel Moellendorf (2005: 2):

> A widely held precept of justice, indeed morality in general, is that we should refrain from foreseeably and avoidably harming others. But if the massive and desperate poor of the world . . . is the foreseeable and avoidable consequence of social conditions shaped and enforced by us, the *advantaged citizens of the affluent countries*, then, according to Pogge, *citizens of the rich countries* are participating in the largest, although not the gravest, crime against humanity in the history of the world. Pogge seeks to defend the thesis that *advantaged citizens of the affluent world harm the world's poor* . . . [emphasis added].

Given the scale of human suffering that will result from climate change, we can find ample support for this view. Pogge's (2005: 34) focus is 'on duties not to harm as well as on duties to avert harms that one's own past conduct may cause in the future [and] on harm we are materially involved in causing rather than all the harm people suffer.' He says that he is motivated to take this approach because duties not to harm other people are more stringent than positive duties to aid people for various reasons. If we apply this to climate change, once again we see that the harm we (the world's affluent people) are causing to others (especially the world's poor people) leaves little alternative but the clear requirement that we change our ways in this respect.

Pogge (2005: 36) is quite emphatic that *affluent citizens of rich countries* are the ones morally responsible for 'immense deprivations' felt by the global poor because they are able to alleviate (cheaply and quite easily) those deprivations, because they are involved in helping to shape and enforce the social institutions causing them and because they benefit from the resulting inequalities. We can probably accept this as an accurate depiction of reality. But Pogge does not go far enough. He says that 'the better-off – we – are *harming* the worse-off insofar as we are upholding a shared institutional order that is *unjust* by foreseeably and avoidably (re)producing radical inequality' (Pogge 2005: 42). At least if we are talking about climate change (but also more generally), the 'we' here must include many or possibly most of the better-off people *everywhere*. Pogge (2005: 45) asserts that one would find it difficult to deny that 'reasonably privileged citizens of the rich democracies share some responsibility for the global institutional order which their governments are shaping and upholding'. He is focusing here on democracies apparently because in such communities people share responsibility for what governments do. But surely this kind of argument cannot be used to deny as well that the politically and economically privileged classes in developing countries, even most non-democracies, support what their governments are doing to perpetuate the international system that is exacerbating climate change and the resulting human-rights violations. China again comes to mind; there public support for the government's economic and trade policies is very strong. Given the huge number of people in China (1.3 billion or more), one might argue that the roughly hundred million most affluent people in China are causing harm to far more people within China alone than do *all the people* in, say, Germany (including poor Germans, of which there are thankfully relatively few). At least we could surmise this. If we add on top of this the greenhouse gas emissions of these kinds of affluent people in all the developing world, there is truly an immense amount of deprivation that will result from the climate change caused by their pollution.

Pogge's argument, under present and predicted environmental circumstances, seems unassailable. But it is missing something that may be especially particular to climate change. What is striking (and typical) about this portrayal of cosmopolitan discourse, from the perspective of climate change, is its focus on advantaged citizens of the *affluent* countries only; it ignores the empirical realities associated with climate change – namely, that *advantaged citizens of the developing, even poor, countries* are also now involved in a massive, foreseeable and avoidable wave of voluntary consumption and energy use that is perpetuating and exacerbating the 'crime against humanity' of which Pogge laments.[4] Pogge could avoid much attention to advantaged people in developing countries because he was not referring to climate change per se (and his audience is advantaged people in developed countries); the international economic injustices he laments were not, at least until recently, sig-nificantly caused by large numbers of affluent people in poor countries. But it is much harder to say that today, unless one accepts that the millions of people in the developing countries who are living better than many people in the West did not willingly embrace, enforce (cf. Pogge 2005: 33) and extend the Western model of development, at least until the global financial meltdown that hit in 2008. (The Chinese mantra, 'to get rich is glorious' (see Harris 2004) is hard to blame on the global economic order!)

In the up-and-coming developing countries, affluent people have choices and are actively involved in perpetuating and extending the global 'order' that seems certain to exacerbate hugely the West's excesses and thereby set climate change running wild. Indeed, some might say that the best opportunity to reform the current global order exists primarily in places like China. If affluent Chinese could avoid the greed and avarice of most Americans, they might set an example for people in the United States and affluent people throughout the devel-oped world. As Debra Satz (2005: 51) asks in her critique of some of Pogge's arguments, 'Is a laid-off American steelworker, for example, really more responsible for global poverty [or, we might add, global warming] than a rich citizen of a poor country?' Pogge's emphasis is also on negative duties – here this would mean not contributing to climate change – which is a good and essential start. However, the affluent people of the world ought to do more; we also ought to aid the poor who suffer from the effects of climate change. As Dale Jamieson has said, the behaviours responsible for climate change 'are part of a lifestyle that is characteristic of the rich but largely foreign to the poor. To a great extent, global environmental change involves the rich inflicting harms on the poor in order to maintain their profligate lifestyles' (Jamieson 1997: 116). This is essentially what Pogge (2002c) is arguing; we ought to help others because we cause them harm. We might say that we ought to act regardless of whether we cause them harm, but we have even more obligation to do so, and to provide aid, if we are indeed causing harm to them.

The deficiency here is the failure to be much more upfront about how 'the better-off' in both developed *and* developing countries 'can improve the circumstances of the worse-off without becoming badly off themselves' (Pogge 2005: 37). Pogge is, of course, aware that 'political and economic "elites" of most poor countries' share the 'immense crime in which we affluent citizens of the rich countries . . . are implicated' (Pogge 2005: 37). Where Pogge seems to stop short, at least if we apply his ideas to climate change, is in not being more explicit about extending his acknowledgement of the role of some people in developing countries to cover the hundreds of millions of people who now fall into the ballooning category of 'economic elites of most poor countries'. If we focus on their part in climate change, it is hard to see how the logic and morals that apply to advantaged people in affluent countries do not also apply fully to advantaged people in poor countries. Again, by acknowledging but under-emphasising the role of affluent people in developing countries, Pogge's arguments applied to climate change face more resistance from the people he most wants to do the right thing – affluent people in affluent countries.

What we should take from this is not a rejection of Pogge's arguments. They identify monumentally important themes that should inform the climate change regime and our political and personal responses to this problem. Rather, we should avoid the advantaged citizen-affluent state myopia and extend these arguments to their logical conclusion: *all* advantaged persons, regardless of whether they live in an affluent state or a poor one, harm the world's poor. From this we can conclude that all those advantaged people, regardless of their state (all things being equal), have certain obligations. This is not only a sounder moral position, but also a good one politically: it removes the incentive of advantaged people in affluent countries to point at advantaged people in developing countries and ask, 'If they don't do anything, why should I?' It is probably wrong to ask such a question (especially if the asking is intended to get oneself off the hook), but one imagines that it goes through people's heads and reinforces the resolve of politicians and policy-makers in developed countries to resist demands that their people and their governments do what Pogge (and global justice) demand. At least in the context of climate change, Pogge's (2002b: 175) focus on affluent citizens of affluent countries creates practical and political obstacles (see Chapter 7). Those citizens will be much less likely to do the decent and humane things that Pogge wants them to do because they can readily see more and more affluent people in developing countries doing exactly what they are being told not to do.

CAPABILITIES OF THE AFFLUENT

One can argue that we have an obligation to aid those in need whom we have not harmed. Peter Singer's arguments regarding aiding the world's poor offer another possible model for bringing *all* affluent people into climate change discourse and policy. He proposes that people who have sufficient money to purchase luxuries should give at least 1 per cent of their earnings to people 'who have trouble getting enough to eat, clean water to drink, shelter from the elements, and basic health care [all of which will be more difficult to get with climate change]. Those who do not meet this standard should be seen as failing to meet their fair share of a global responsibility, and therefore as doing something that is seriously morally wrong' (Singer 2004: 194). Here the notion is that those with the ability to take action, in particular to aid those suffering from climate change, ought to do so, regardless of whether they accept or recognise some cause-and-effect relationship. Importantly, Singer points to rich people in poor states: 'Not all residents in rich countries have income to spare, after meeting their basic needs; but . . . there are hundreds of millions of rich people who live in poor countries, and they could and should give too. We could advocate that *everyone with income to spare*, after meeting their family's basic needs, should contribute [a small amount] of their income to organizations working to help the world's poorest people' (Singer 2004: 193, emphasis added). This is the kind of explicit discussion that we need much more of, alongside discussions of interstate obligations in the context of climate change.

Jamieson also believes that causing harm may not be as important for determining moral responsibility in the case of climate change as is the *ability to benefit or prevent harm*: 'those who are in a position to prevent or mitigate climate change are responsible for doing so regardless of their causal contributions' (Jamieson 1997: 117). This follows from Singer's (1996) seminal principle – if it is in our power to prevent something very bad from happening, without thereby sacrificing anything that is morally significant, we have a moral duty to do it – and means that, if we are able to do so, we should 'seek to make things better by trying to do no harm' (Jamieson 2005: 167). So here we see a basis for affluent people everywhere to act now to limit their greenhouse gas emissions; their obligation does not depend on anticipating future harm they may cause to others. Jamieson (1997: 118) argues that those who are able to do so 'should seek to stabilize climate, and they should also do what they can to help those who are most vulnerable to the change that may already be occurring'. A conclusion we can draw is that affluent individuals *everywhere* ought to restrain their greenhouse gas pollution and aid those suffering now from climate change.

Elliott (2006: 355) makes another argument that is essentially divorced from a direct cause-and-effect relationship. She argues that an ethical prescription responding to environmental harm requires that we do more than not commit harmful acts affecting others. Environmental harm is more than the obvious harm caused by environmental degradation. Because it is also 'the harm caused to people by environmental injustice, the cosmopolitan duty to avoid environmental harm demands that one avoid or take reasonable precautions to avoid causing environmental injustice as well . . . We should act individually or collectively to prevent such harm and injustice from occurring even if we are not the specific instigators of such harm' (Elliott 2006: 355). Indeed, Elliott suggests that there may also be a positive duty to assist people who are harmed by environmental degradation to overcome that harm. At the very least, there is a 'meritorious duty' for those with the ability to provide assistance to victims of environmental injustice to do so, following Andrew Linklater's (1999: 476) invocation that 'we have obligations to help the poor overcome the effects of inequalities, even if we have had no part in creating them' (quoted in Elliott 2006: 355).

The conclusion that a cosmopolitan viewpoint reveals is that people (not just governments) are at least obliged, especially if they are affluent (but not if they are poor), regardless of their nationality or where they may reside, to act in ways that do not undermine others' environmental rights, and, especially if doing so is not a great hardship, they ought to aid those who suffer from a lack of those rights.

THE POLLUTER-PAYS PRINCIPLE

Environmental globalisation means that we all increasingly affect each other regardless of where we live. However, the causes of our pollution, notably through our use of fossil fuel energy and other activities leading to emissions of greenhouse gases, are now being greatly supplemented in pockets of affluence in poor countries among new consumers who are joining the majority of people in the developed world in polluting the atmosphere. The consequences of this pollution are experienced even in pockets of poor people in the affluent countries who join the majority of people in poor countries adversely affected by climate change (for example, many of Hurricane Katrina's poorest victims). According to Sachs (2002: 40), the 'victims of the degradation of living systems live predominantly in the South, or more precisely, are typically part of the majority beyond the corporate-driven consumer class in North and East *and* South' (emphasis added). Hermann Ott and Sachs (2002: 166) point out that 'assuming that states are relatively homogenous internally shields the fact that huge disparities among social classes exist within states', with large groups of affluent people in developing countries (for example, Brazil, China, India, Mexico) consuming as many goods, and using as much energy, as their affluent fellows in the developed countries, implying levels of consumption that are five to ten times their countries' averages. Consequently, 'if the "polluter pays" principle were applied not to states, but to members of the global middle class, then most of the Southern middle classes would have to accept reduction commitments already today' (Ott and Sachs 2002: 166).

If typical Britons or Italians are equivalent to upper-middle-class people in the developing world, it follows that we can at least say that everyone who is in the upper middle class (or higher) in the developing world has about the same moral obligation to act as do typical Britons and Italians. It is safe to say that the number of people who fit this category in the developing world is at least in the tens of millions, and probably in the hundreds of millions. That translates into a tremendous amount of greenhouse gas pollution. Vanderheiden (2008) is part of a small group of scholars looking at climate change from a perspective of broadly global justice (rather than only international justice) in a way that captures these new consumers. He points out that 'all the world's nations *and citizens* are stakeholders and participants in what must be a fully global effort if the worst effects of what could be humanity's most pressing global threat are to be avoided' (Vanderheiden 2008: 81–2, emphasis added). Thus, from this cosmopolitan perspective, the solution is not in simplistic and unrealistic classifications of 'Annex I' and 'non-Annex I' countries (as most developed and developing countries are respectively classified in the climate change agreements), where all citizens carry labels of rich or poor based on nationality and regardless of their real wealth and well-being. Cosmopolitan justice demands that we explicitly recognise the contribution of the world's new consumers to climate change, and that we bring them into the ethical debate and the practical solutions, rather than mostly ignoring them as international deliberations and instruments do at present.

The familiar polluter-pays principle ought to apply: each individual who pollutes is obliged to act and to aid if he is affluent, regardless of whether he lives in a rich or a poor state. Thus, all things being equal, a poor person (measured by some reasonable standard of purchasing power parity) in Germany may be less obligated to act on climate change than is an affluent person in, say, China or Chile, if the former pollutes less. There is even an argument to be made for going a bit more easily on affluent people in the industrialised world because they did not know until quite recently that they were doing harm to the global climate and because they are stuck in economic structures, and with infrastructure, premised on the use of fossil fuels.[5] As Ott and Sachs (2002: 169) argue, 'industrialised countries do not start from scratch, but have locked themselves into a fossil-based infrastructure that cannot be dismantled in the short and medium term. This may entitle them to a "bonus" for a first mover disadvantage.' And not all those people have benefited greatly from the fossil-fuel-based economies in which they live (compare, for example, the people of rich New Orleans, who suffered relatively little from Hurricane Katrina, to those people of poor New Orleans, who suffered much more and are still suffering).

Thus, arguably, the newly affluent people in developing countries may have greater moral obligation than many people in affluent countries to keep their consumption down – or to bring it down if they have already started to consume more than they need – because they are not yet locked into a highly consumptive lifestyle, especially if they are not yet 'trapped' in a national economy predicated on unnecessary consumption and pollution. This suggests that affluent Chinese may have more obligation than do many or most affluent Canadians, because the latter are saddled with infrastructure and long-standing habits that were created before they knew that climate change was a problem. Educated Chinese people (and their government, media, and so on) know better (or should) because scientific knowledge about climate change, and associated high-profile international diplomacy, have coincided with China's economic rise.[6] As Vanderheiden (2008: 188) argues, we cannot simply let people claim ignorance. Doing so would provide an incentive for wilful ignorance to avoid problems and would encourage governments to cultivate ignorance in order to avoid responsibility to regulate pollution (as happened with the very effective 'climate-sceptics' campaign in the United States). At the very least, this reinforces the point that the new consumers cannot claim that they have no moral obligations to respond to climate change.

The normal response to this sort of argument is that the poor states are not responsible for climate change now and in the near future, nor are their citizens. This is mostly true if we ignore affluent people in poor countries in the past, which we can do because, in practical terms, they mattered relatively little. However, Garret Hardin (1968: 1245) presciently noted decades ago that analysis of pollution problems 'uncovers a not generally recognized principle, namely: the morality of an act is a function of the state of the system at the time it is performed'.

Vanderheiden (2008: 163–4) reminds us of what really happens: 'some act that once was entirely benign, and would still be given smaller global population, *became* harmful once some threshold level of population or emissions was exceeded. Literally, the wrongness of some act may depend on how many other people are able to benignly commit that same act . . .'. This challenges conventional ethical wisdom and again highlights the role of the new consumers.

John Marburger (2008: 51) believes that this assumption of situational ethics 'speaks to the morality of censuring behaviour today (the generation of greenhouse gases) that was acceptable in the past', concluding that countries that have enjoyed the fruits of the Industrial Revolution bear no special moral burdens, although he believes those countries still ought to help developing countries bear burdens of the present. Importantly, he concludes that developing countries have not inherited a 'right to pollute', because the 'situation has changed, the consequences are known, and the impacts are ultimately universal. Countries that are aware of the possibility of adverse impacts on their populations are morally correct to seek changes in behaviour in all other countries that pollute the common resource sink, regardless of history' (Marburger 2008: 51). While Marburger seems to be letting the developed countries off the hook too easily, there is an important point implicit in his argument: even rich states at least have a right to seek changes in the behaviour of *some actors* within *all* other countries, if they pollute the atmospheric commons, especially when the effects are felt (or will be) by the least advantaged people within the complaining states.

In his discussion of contributions to global poverty, Pogge (2005: 48) patently rejects the argument that '*we* should not be required to stop *our* contribution until *they* are ready to stop *theirs*', because if such an argument were right 'then it would be permissible for two parties together [such as two upstream factories releasing chemicals into a river (Pogge's example, which is highly analogous for our discussion of climate change)] to bring about as much harm as they like, each of them pointing out that *it* has no obligation to stop so long as the other continues'. Given the harm that would result from such logic, Pogge concludes that 'one may not refuse to bear the opportunity cost of ceasing to harm others on the grounds that others similarly placed continue their harming' (Pogge 2005: 53). Of course this sounds very much like common wisdom: two wrongs do not make a right. More generally, however, the argument that greenhouse gas emissions in the past must be dealt with before requiring common action today is ethically problematic because it can be made perpetually: 'However much it may deplore the effects of the previous generations' decisions on it, each generation will face the same decision situation with respect to generations later than it' (Gardiner 2001: 404). The only logic to the historical argument for inaction is a narrowly self-interested one of states concerned with the here and now, and those people who benefit from the status quo.

Consequently, history may judge affluent Chinese (for example) even more harshly than many people in North America, Australia and Europe because the former had (and still have) a choice about whether to jump on the consumption bandwagon. Of course, the Chinese government is responsible and complicit. It is sending people to wave the Chinese flag in Earth's orbit while millions of its own citizens live in squalor. At the same time (to continue the theme of outer space), it is all but certain that among the first space tourists will be wealthy Chinese who have themselves conspired in the government's environmentally harmful growth. While average per-capita greenhouse gas emissions in China are well below the averages for the world and especially the developed world – but above those for the whole developing world – a burgeoning middle and upper class is hiding behind this average. It can do this for only so long. 'Luxury' emissions of greenhouse gases that result from the voluntary consumption choices of China's affluent people, most perversely the country's new super-rich elites, are multiples of both Chinese and global averages, and indeed well above the averages of the major polluting states of the developed world. Similarly, in the rest of the emerging developing world, the growing middle classes, while still only minorities within their own countries, are consuming above their weight. As the spending on luxury goods and travel explodes in China, India and many other developing countries, this practice of hiding behind their compatriots' overall poverty becomes more and more perverse, not least because most of those who will suffer the most from greenhouse gas emissions from these newly affluent people will be people living in those same countries.[7]

Some people will strongly disagree with the idea that we need to focus more on individual obligations. For example, in questioning the relevance of Singer's utilitarianism (1996) in the case of climate change, Mathew Paterson (1996: 190) complains that 'Singer's version places the location of obligation also at the level of the individual, rather than at the level of social and political institutions. Therefore, while this might be a guide to action for individuals (for example, at the crude level, "stop using your car to help those in small island states"), it is not clear how political institutions should respond', because it is difficult for them to decide how to do so (for example, discourage the use of cars or increase their efficiency), they have competing obligations and 'the relationship between the intention of the action and its result is much less clear than in the case of individuals'. However, this seems to be a strong argument *for* individual action: there is a clearer 'relationship between the intention of the action and its result' than would be the case if we talk about institutions, which of course themselves do not pollute at all. A lack of this clarity is hardly an excuse for governments to do nothing, but it need not be one or the other: both individuals and states (and other social and political actors) can and should act. Paterson has other complaints about the utilitarian-individual approach. He says that it 'may well be impossible to apply at the global level. The complexities involved in global warming lead to it being impossible to ascertain what might improve the general level of welfare' and, very importantly, 'global warming throws up great questions concerning the *meaning* of human welfare. Do we value material goods and economic growth over risks to do with climate change impacts and so on?' (Paterson 1996: 190). Given the scale of harm and suffering that is on the horizon, the answer becomes less and less difficult to see. More generally, these criticisms, while worthy of consideration, seem to be (unintentional) foils that deflect obligation away from (affluent) persons and onto (affluent) states. But both can be ethically obliged to respond to climate change.

ABSOLVING AND IGNORING THE NEW CONSUMERS
Mahathir Mohamad, while Malaysian Prime Minister, was one of those developing-country politicians who strongly criticised the notion that the world's poor countries should cut their pollution. Mahathir (1998: 325–6) argued that what the *rich* do is what matters:

> We know that 25 percent of the world population who are rich consume 85 percent of its wealth and produce 90 percent of its waste. Mathematically speaking, if the rich reduce their wasteful consumption by 25 percent, worldwide pollution will be reduced by 22.5 percent. But if the poor 75 percent reduce consumption totally and disappear from this earth altogether the reduction in pollution will only be by 10 percent. It is what the rich do that counts, not what the poor do, however much they do it. That is why it is imperative that the rich change their life-styles. A change in the life-styles of the poor only, apart from being unfair, is quite unproductive environment-wise. But the rich talk of the sovereignty of the consumers and their right to their life-styles. The rich will not accept a progressive and meaningful cutback in their emissions of carbon dioxide and other greenhouse gases because it will be a cost to them and retard their progress. Yet they expect the poor people of the developing countries to stifle even their minute growth as if it will cost them nothing.

It is very hard to disagree with this logic from both practical and ethical perspectives. What Mahathir says is unjust is indeed unjust. The problem is that he and other developing-country politicians and state-spersons, and their advocates, have used this argument in the context of *international* negotiations on climate change. He used it, and rightly so, to help developing *states* avoid requirements that they limit their green-house gas emissions and other pollution at the expense of poverty eradication. Meanwhile, the 'sovereignty' of rich people in Malaysia and other developing countries to consume at will has been defended and even strongly encouraged.[8] So really Mahathir seems to have been against people in the developed countries polluting the planet, not his rich co-nationals doing exactly the same thing.

Let me return to a brief quotation from Chapter 4. According to Benito Muller's assessment (2002: 2) of the developing countries' own view of climate justice, 'the paramount inequity is one of human victims and human culprits'. If we take this as an accurate assessment of what the developing countries have been arguing for, then even they are making a cosmopolitan argument that focuses on human beings (not on states) and their responsibility for causing harm to other human beings (not to other states). But the developing country governments seem to want this interpretation of climate justice (consistent with Mahathir's argument) to apply only to the people of the *developed* countries. The very good argument that Mahathir makes thus becomes a *distraction* from universal obligations that are increasingly part of the climate change problem and potential solutions to it. The intent seems to be to persuade people in affluent states to cut their pollution and to aid poor *states*, not to have *all* affluent people do so – let alone directly aid poor *people* in developing-countries. This way of thinking and arguing, by diverting our attention to the developed states' (and their citizens') duties and to developing states' rights, may have contributed to the failure of the climate change regime. If the rich in the poor countries were seen to be behaving responsibly, it would be much harder for some people in the West to make the unjust argument that the developing countries must act robustly before they will agree to do so, or even to argue that all states must move together. Really what is needed, ethically, is for all affluent people to move together. This would result in movement everywhere, albeit much more in the rich states. In short, common but differentiated responsibility among *states* is no excuse for affluent people in developing countries to delay cutting or limiting their greenhouse gas emissions alongside affluent people in developed countries. While responsibilities of states are legitimately differentiated, those of most affluent people are not.

One very simple (and I think unassailable, albeit conveniently ignored) response to people everywhere who condemn (as they should) the environmental rapacity of the United States and of most Americans is this: 'if you find America's or some other country's behaviour morally outrageous, and the principles operative in your thinking this apply to you too, then you should find your own behaviour morally outrageous as well' (Garvey 2008: 142). That a rich person in Mumbai or Shanghai could reject this logic with a straight face is one of the greatest obstacles to protecting the atmosphere in the future. The arguments now being used by many rapidly developing-country governments (rightly so) to beat up the Americans will one day be used against those same governments. If the historical emissions of Americans are worthy of blame (as they certainly are, at least since the last decade of the twentieth century), then emissions resulting from affluence in, say, Hong Kong and Iran (and similar regions and states), whose governments and citizens are immune to legal obligations in the context of the climate change regime, are also worthy of the same blame and require the same response. To reach a significantly different conclusion would be hypo-critical.

An example of how the rich in developing countries seem to get off the hook, if not morally than perceptually, is commonly found in ethical discourse. For example, Brock (2007: 136) writes that 'attention to the global needy is more of a requirement than is commonly appreciated by the affluent of developed countries'. Now look at that statement. Either it seems to suggest that the affluent of develop*ing* countries are more attentive to the global needy than are the affluent of developed coun-tries, or it suggests (which I assume is what Brock meant to say even if she did not mean it quite so literally) that affluent people *in developed countries* ought to be more attentive – implying, of course, that the affluent in developing countries need not heighten their attentiveness to the global needy, least of all to needy people living in, say, Madrid. Similarly, Cecile Fabre (2007: 142) points to political philosophers (Charles Beitz, Charles Barry and Thomas Pogge) who argue that 'rich members of rich countries should distribute part of their wealth to poor members of poor countries'. Even if we agree with this assertion, as I think we should, it is rather explicit in ignoring the question of whether rich members of poor countries ought also to distribute part of their wealth to poor members of poor countries, to say nothing of whether they should do the same for poor members of countries that are not poor.

Another example is found in Darrel Moellendorf's cosmopolitan defence (2002: 98) of the argument that 'persons in the developed countries should subsidize the costs of responding to the effects of global warming for persons in developing countries'. On first reading this is certainly justifiable. But, on second reading, one is left asking whether Moellendorf really means to say that poor people in San Diego should be subsidising persons in developing countries, and whether he means to exclude affluent, capable people in developing countries. One assumes that he would answer 'no' to both queries. What is significant is that he, like most people who comment on these issues, including most other cosmopolitans, leaves this unsaid. The discourse gives weight to the notion that what really matters for cosmopolitan climate justice is rich people in rich countries, full stop, even as the weight of the causal impact shifts quite significantly in the direction of rich people in poor countries. This is partly an empirical issue; the new consumers are new, after all. But it may also be evidence of the power of international doctrine to introduce bias even into cosmopolitan arguments. What is left unsaid is the fundamental moral cosmopolitan argument: everyone is equal, and therefore every affluent person who pollutes the atmosphere is equally responsible (*ceteris paribus*) for acting to deal with it, whether that be by reducing atmospheric pollution, aiding those who suffer from its effects or taking other required steps. As Simon Caney (2006b: para. 12) argues, 'persons are entitled to the protection of their fundamental interests from the harmful effects of global climate change and it is unjust for other persons to act in ways which would leave people's fundamental interests at risk from the changing climate'.

Many of the world's affluent and privileged people will respond that climate change is really not their fault, that their personal contribution to climate change is tiny. This is largely true, but of course if everyone who is affluent thinks this way, and behaves accordingly, the aggregate pollution can be enormous, especially as the number of affluent people grows. This practical truth is belied by the immorality of avoiding responsibility. As Pogge (2002b: 170) argues with regard to global poverty, 'even a very small fraction of responsibility for a very large harm can be quite large in absolute terms . . .'. In the case of climate change, the affluent consume disproportionately more and in so doing emit disproportionately more greenhouse gases than do the poor. Pogge (2002b: 192) notes that, 'even if each privileged person typically bears only one billionth of the moral responsibility for the avoidable underfulfillment of human rights caused by the existing global order, each of us would still be responsible for significant harm'. He also acknowledges that 'nearly every privileged person might say that she bears no responsibility at all because she alone is powerless to bring about a reform of the global order' (Pogge 2002b: 170). However, Pogge (2002b: 170) points out that this 'is an implausible line of argument, entailing as it does that each participant in a massacre is innocent, provided any persons killed would have been killed by others, had he abstained'. Vanderheiden (2008: 166) makes this point with regard to climate change in particular: 'Isolated individual contributions to larger aggregate problems may appear to be trivial, yet the countless occurrences of such seemingly trivial acts together add up to quite serious harms'. Similarly, Garvey (2008) points to the refrain that people often make: that the harm done by the United States is so great, why should my tiny bit of harm matter? Garvey's response: 'It might be that both my failure and America's failure are the same sort of wrongs, the same in kind, different only in magnitude. It is possible to think that my failure to do something about my high-carbon lifestyle really is morally outrageous' (Garvey 2008: 142).

One problem with international/interstate conceptions of climate justice is that they can make people lazy; they push duties and responsibilities onto governments. People can say, 'I pay taxes and follow regulations. I have done my duty'. This applies to affluent people in rich and poor countries alike, neither of which should be allowed to avoid responsibility if their governments have failed to implement policies necessary to push them to behave according to reasonable requirements of global justice. Nor should they be allowed to shirk their duties to limit their greenhouse gas emissions and, in the case of people in democratic societies, to support politicians and policies that will address climate change and its consequences – and *not* to support politicians and policies that do otherwise. In Garvey's view (2008: 143), 'one of the conclusions we are avoiding, perhaps above all others, is a personal conclusion: I ought to change my comfy life'. While it might be hard for affluent people to consume less, we can at least work very hard towards that end (as Pogge suggests in the case of human rights) and to contribute to organisations working to 'help prevent or mitigate some of the harms caused by the global order' (Pogge 2002b: 171).

It really is no longer acceptable for capable people, especially affluent capable people, to continue to do nothing or, worse, to embrace the kind of lifestyle that has largely got the world into its current ecological mess. Given the importance of individual contributions to climate change, if international environmental treaties and regimes are to be effective, it is time for them to include explicit obligations and duties for all affluent individuals to implement. As Dower (2007: 8) warns, 'the world situation is such that not only is it right or one's duty to do these things, but also if we do not generally do these things then we are storing up trouble on a global scale'.

Cosmopolitans like Singer (2004: 152) rightly complain about the preoccupation with national boundaries and a charity-begins-at-home attitude, particularly in the United States. They do this to point out the moral obligation to aid the poor abroad irrespective of state borders. It is much less common to hear arguments that affluent people in poorer countries share the obligation towards the world's destitute that obtains among affluent people in affluent countries. Nevertheless, if one accepts the cosmopolitan ethic (and logic), one is left with the rather obvious but rarely considered conclusion that the affluent in poor countries have the same obligations as affluent people in affluent countries to restrain their consumption and pollution. When there were relatively few affluent people in the developing world, as was the case until quite recently, we could overlook the new consumers' practical impact on climate change and let them free ride on the limited obligations of their states, much as the rich have always been free-riders. However, with the numbers of affluent and even wealthy people in developing countries expanding so much and so quickly, the ethical and practical importance for them taking responsibility and acting accordingly is no longer something we can ignore – at least not if we want to combat climate change robustly and address the injustices experienced by those people and communities most affected by it.

We can remain sympathetic to those who point out that per-capita emissions identify the United States and a number of other industrialised countries as far and away the world's greatest state polluters: most people in those rich countries pollute heavily, even grotesquely, and the United States was until recently the largest national source of greenhouse gas pollution. Pollution from the United States, and those other rich countries, is very unfair. Developing-country governments rightly focus on the unjust *luxury* emissions of the rich versus the *survival* (or subsistence) emissions of the poor (cf. Shue 1993), noting that the former ought not to come at the expense of the latter. But they are talking about the luxury emissions of the rich in rich countries. They almost never talk about the luxury emissions of the rich in *poor* countries. It is as though those people do not exist in ethical terms. To be sure, their practical importance has been much less than that of people in the developed world. But that is changing very rapidly. While we do not say it outright at present, our silence conveys the message that the luxury emissions of the affluent in poor countries are in some sense in the same ethical category as survival emissions of poor people in poor countries. Cosmopolitanism by definition rejects the negligence that leads to such a perverse result; ethical obligations (and rights) exist *regardless of nationality*. Consequently, fat cats in Shanghai have just as much of an ethical obligation to reduce pollution as do fat cats in

London. As the number of the former grows, we ought to acknowledge the importance of this in practical terms, in the process being more ethically consistent.

CONCLUSION

Most people would agree that affluent people in the rich countries should bear at least some ethical responsibility for harm they do to the world's poor; this is arguably part of civic responsibility in most developed democracies (Satz 2005: 50). Many people would also agree that the affluent classes of rich countries ought to give aid to the poor world simply because that aid is needed. The rise of the world's new consumers does not eliminate the ethical obligations of developed countries to aid poor ones, nor does it alter the obligations of affluent people in the developed countries to do what is right. The question is whether affluent people everywhere have the same responsibilities, and whether we are willing to acknowledge this more than we do at present. A robust cosmopolitanism captures affluent consumers in the developed states *and* the new consumers as well. One way to interpret this is that a cosmopolitan viewpoint is more practical than international doctrine because it encompasses vastly more of the causes of climate change.

As Sachs and Santarius (2007: 158) argue, justice requires 'a retreat from over-appropriation of the environment' by the world's 'transnational consumer class'. Capable people in both developed and developing countries ought to follow Aldo Leopold's dictum ([1949] 1968: 224): 'A thing is right when it tends to preserve the integrity, stability and beauty of the biotic community. It is wrong when it tends otherwise.' As Elliott (2006: 355) puts it, 'the cosmopolitan commitment to minimise environmental harm may well be seen as something approaching a perfect duty, that is, we should minimise environmental harm as a general good, even if we do not know specifically who might be harmed or protected from our actions'. More generally, an environmental ethic 'is if nothing else an ethic of interdependence, and will not countenance the excuse "I don't intend to help spoil the environment – all I intend to do is get to my meeting ten minutes earlier by private car than by public transport"' (Dower 2007: 179).

In a discussion of distributive justice, Stanley Hoffman (1981: 144) describes two versions of what might constitute global justice: the 'classical', which is concerned about states, and the 'radical', which is concerned about people. One problem with the first version, which demands that rich states provide aid to poor states, is that it 'really amounts to a reinforcement of the state system' (Hoffman 1981: 147). If obligation is between states, there is no assurance that individuals in the recipient communities will benefit. The radical version, notably an 'extreme-radical' variant concerned about the causes of unequal distributions of wealth among individual persons in most states as well as among states themselves, is significant here because 'it blurs the distinction between states and individuals, and even deems the distinction illegitimate. It states that problems of distributive justice in international affairs are problems of duties to individuals, and it suggests that the problems of state inequality . . . are either irrelevant or subordinate' (Hoffman 1981: 148). Hoffman identifies a 'minimalist' position – there is no moral obligation to individuals because there is no community of mankind, for example – and a 'maximalist' position, reflected in the ideas of Charles Beitz (1979b), which asserts that 'our obligation concerning justice is universal, that despite the existence of separate states and nations we have a duty to all mankind' (Hoffman 1981: 153), because, among other reasons, there is a kind of global community and that is what we would choose if (borrowing from Rawls 1971 (cf. Brock 2007)) we were in an original position, behind a veil of ignorance, not knowing our nationality and so forth. To invoke the words of Hoffman (1981: 153) once again, 'to put it bluntly, our obligation of justice toward the Bantus is exactly the same as our obligation toward our immediate neighbors'.

Hoffman rejects both of these ideal positions, ending up 'with the philosophically untidy and politically elastic notion that the scope of our obligation to individuals in other societies varies in time and in space' (Hoffman 1981 157).[9] What is germane here is that Hoffman is advocating some middle ground between focusing on the obligations (and especially the needs) of individuals and focussing on those of states. Preoccupation with the latter is neither ethical nor practically reasonable. Having said that, climate change provides support for radical-maximalist arguments for justice by placing everyone, everywhere, in a situation of mutual dependency. There is no American or Chinese climate system (as distinguished from weather systems); there is only one atmosphere, and every person contributes to changes in global climate, albeit with varying effects in different places, regardless of where he is located. Obligations – for states *and* for people – arise from this circumstance.

The current situation, with affluent persons harming others, notably the poor and the weak of the future, seems blatantly unjust. On what basis can we say it is unjust? Not on the basis of *international* justice, which does not ascribe obligations to individuals per se. We need an alternative justification, which can come from cosmopolitan conceptions of justice suggesting that *individuals* are citizens of 'one world' (cf. Singer 2004). It is hard to refute the empirical argument often made by cosmopolitans: borders do not matter the way they did in the past (Attfield 2003: 159). As Beitz (1979b: 176) put it, 'when, as now, national boundaries do not set off discrete, self-sufficient societies, we may not regard them as morally decisive features of the earth's social geography. For purposes of moral choice, we must instead regard the world from the perspective of an original position from which matters of national citizenship are excluded by an extended veil of ignorance'.[10] With the rise of the new consumers in the developing world, the sources of climate change are now widely distributed across the globe. Thus a realistic and practical viewpoint is needed alongside the cosmopolitan moral viewpoint. With this in mind, in Chapter 7 I look at how the connection between cosmopolitan justice and climate change might be made part and parcel of the climate change regime and some of the related international negotiations and national policies. In Chapter 8 I extend this to 'personal policy' as well.

NOTES

1. The data used by Myers and Kent (2003) are from 2000, meaning that they do not count many millions of people who have became new consumers since then. They also do not count about 140 million new consumers in many smaller developing countries.
2. This level of 'significant discretionary consumer spending' is set by Garner (2006: 73) at $5,000. In comparison, the number of Americans with income over this amount was 108 million in 2003 (96.6 per cent of all households).
3. Modifying Shue's words (1992: 397): 'whatever justice may positively require, it does not permit poor nations to be told to sell their blankets in order that rich nations may keep their jewelry.'
4. Pogge recognises the growing environmental impact of developing countries' economic growth on the world, but he discounts the new consumers who contribute (avoidably) to that impact (cf. Pogge 2008: 218).
5. On responsibility for past emissions, see Gardiner (2004: 578–83). On ignorance of the impacts of climate change, see Caney (2005a: 761–2).
6. China's shift to capitalism began in earnest about 1980, which was around the time that climate change science started to become prominent.
7. For comparison, purchasing power parity (PPP) in the USA was roughly seven times that in China in 2004 (Central Intelligence Agency 2005). But this is across the entire Chinese population, which includes hundreds of millions of people earning very much less than the national PPP average. The obligations discussed here apply not to those people, but instead to the affluent and wealthy in China – and everywhere else, regardless of the size of their national economy, its total PPP figure or per-capita PPP.

8. By way of example, Mahathir was a big advocate of car production in Malaysia.
9. This leads Hoffman (1981: 158) to say that we owe aid to *states* to assist poor *individuals* within them, except 'insofar as there are violations of the most elementary human rights of other individuals', so he is still very much affirming sovereignty.
10. Beitz is invoking Rawl's 'original position' and 'veil of ignorance' (see Chapters 2 and 5).

CHAPTER 7

COSMOPOLITAN DIPLOMACY AND CLIMATE POLICY

If any issue demands a cosmopolitan response, climate change is it. It is a global problem with individual causes and consequences. Because cosmopolitanism is concerned with individuals, it can help the world solve the dilemma of otherwise rational economic development within states, and international cooperation among them, that is tragically not preventing – and indeed contributing to – severe harm to the planet's climate system. Cosmopolitan justice addresses the disconnection between the lack of any legal obligation of many millions of affluent people beyond the scope of the climate change agreements – including the affluent in developing countries – to cut their greenhouse gas pollution, and their ethical responsibility to cut pollution alongside affluent people living in the few rich states that have agreed to binding national obligations in the context of the climate change regime. Cosmopolitan justice demands that we explicitly recognise this reality rather than ignore it in the international legal instruments on climate change. Implementing cosmopolitan justice here means that obligations to act on climate change, and to aid people harmed by it, apply to all affluent individuals *regardless of where they live*.

This points to a corollary (or supplement) to prevailing applications of international justice to climate change: a way forward for climate justice that acknowledges the responsibilities and duties of developed states while also explicitly acknowledging and acting upon the responsibilities of all affluent people, regardless of nationality, as global citizens. This cosmopolitan corollary is an alternative to the status quo climate change regime, premised as it is on the rights and duties of states while ignoring the rights and duties of too many people. The corollary is more principled, more practical and indeed more politically

156

viable than the current doctrine and norms of international environmental justice applied to climate change.

In this chapter I continue to query the most common cosmopolitan
arguments about global justice and describe an alternative to the purely
international response to climate change, in the process proposing that
cosmopolitan aims be incorporated as *objectives* of climate change
diplomacy and policy. I describe some of the skeletal features of a
cosmopolitan corollary to state-centric policy responses to climate
change. I suggest some of the ways by which we might reconcile the
failure of the climate change regime to recognise the rights and duties of
persons with the pressing need to bring all capable people, and especially
affluent people – including the world's new consumers – into the picture.
A cosmopolitan corollary to international doctrine recognises everyone's practical and ethical importance. In so doing it offers all actors,
including states, a more politically viable approach to solving climate
change. An objective here is to start outlining how an alternative to
international doctrine might be implemented. I do not claim that these
are the only or even the best ways of actualising cosmopolitan justice in
this context.

The burden of determining how to implement climate diplomacy
must first be preceded by an appreciation of the importance of new
thinking, discourse and policy that are highly sensitive to cosmopolitan
objectives. By associating the wealth and behaviours – and the pollution
– of individuals and classes of people with ethical diplomatic arguments,
international agreements and the domestic implementation of those
agreements, governments of both developed and developing states can
escape the ongoing blame game in which poor countries blame rich ones
for the problem so far, and rich states blame poor ones for the problem
to come – with both refusing sufficiently to obligate even their affluent
citizens to do all that is necessary and just. A new kind of climate
diplomacy premised on cosmopolitanism allows major developing-
country governments simultaneously to assert and to defend their
well-justified arguments rejecting *national* obligations related to climate
change while also acknowledging and regulating growing wealth and
pollution among a significant segment of their populations. Because of
its universal application – because people in developed countries will see
new consumers in the developing world taking action – the cosmopolitan corollary in turn can help to neutralise the understandable (if
unjust) political reticence of most developed-country governments to
live up to their states', and their own affluent citizens', obligations
finally to undertake the major cuts in greenhouse gas emissions that
the Earth requires. The implication of the cosmopolitan corollary to

international justice is an escape from the curse of Westphalian norms. Applied to climate change, cosmopolitan justice has the potential to define a pathway whereby *all* countries, both developed and developing, can participate fully in the climate change regime *without* making any demands on the world's poor – indeed, while aiding them in new ways.

THE COSMOPOLITAN COROLLARY TO INTERNATIONAL DOCTRINE

The climate change regime has failed, at least in part, because it is premised upon a doctrine of states' rights and interests, resulting in neglect of many millions of consumers driving much of the problem. By essentially neglecting these people in their calculations, or at best considering them only by implication, governments have ignored a growing part of the problem of global environmental change (cf. Elliott 2006: 353). From the perspective of the climate change regime, it is as though the millions of affluent people in developing countries have no moral, practical or legal significance. This is arguably immoral, definitely impractical given the growing contribution to global warming of these millions, and politically problematic because the poor and not-so-poor in rich states see the unfairness of it and ask why they should give up cherished Western lifestyles while millions of affluent people in developing countries adopt them.

INSTITUTIONALISING COSMOPOLITANISM IN A WORLD OF STATES

Patrick Hayden (2005: 132) is critical of myopic international doctrine: 'Not only does an overly statist international system make it incredibly difficult to create the ecological norms and consensus needed for a genuinely global response to environmental issues, but also to implement global agreements and guide active and effective compliance.' Hayden does not exclude a role for states, but rather recognises that an obsession with them can be counterproductive. Overemphasis on community and national autonomy may be something that should be abandoned over time, but for now state communities dominate (cf. Brown 1992: 170–1). States are potentially equipped to deal with global environmental problems, and the world is de facto divided into communities that derive much of their identity as states. Justice ought to be aimed at individuals, but states are the mediators that act to achieve it. 'In other words,' to quote Luigi Bonanate (1995: 118), 'justice is an interindividual fact that is regulated by states'. Benito Muller's assessment (2002: 2) of the divide between rich and poor countries in the context of climate justice leads him to conclude that 'at the *decision-making* level, human impacts and their differentiated causal responsibilities must be fully acknowledged and taken into account' in the climate change negotiations among states, but 'the lesson at the level of *policy analysis* must be to put much greater effort into thinking of innovative ways in which these human impact burdens could be distributed'. This seems to demand a cosmopolitan approach, given the declared need to consider the role of human beings, both their contributions to climate change and the effect that it has on them, and more generally their responsibilities and rights.

Thus one might posit that, *especially in the context of climate change*, cosmopolitanism is more realistic than communitarian state-centric approaches to solving global problems. Because cosmopolitanism is premised on the rights and interests of persons, it reveals the true locus of pollution causing climate change and the profound consequences of this pollution for billions of people. But, as we have seen, the way that cosmopolitanism is normally applied to the problem usually falls short, because it focuses on only about half of the people causing it – affluent people in developed states – while ignoring affluent people in developing states who are rapidly catching up with their developed-country counterparts in their power to consume and pollute. Furthermore, most cosmopolitan responses routinely revert to state-centric prescriptions, with the usual argument being something like this: affluent people in the developed countries pollute so much, therefore their *states* must reduce national greenhouse gas emissions and aid developing *states*, and possibly compensate them as well. How can we adapt these state-oriented conceptions and practices of climate justice, as well as most of the cosmopolitan alternatives that tend to surrender to state domination as well? Doing so is not at all easy, which is one reason why many cosmopolitans argue the way that they do. Even when they desire institutions and policies based on cosmopolitan morality, they are realistic and recognise that states and the communities they often engender dominate the world and, for very practical reasons, have to be major actors in finding solutions to climate change. From this thinking it is then normally concluded that, if someone or something is to be blamed, it must be the states where most of the responsible and capable people reside.

Edward Page (2008: 570) points out that Simon Caney's hybrid account of climate justice, which bases responsibility on both causal responsibility and ability to pay, locates the source of justice in the interests of persons. We are reminded, then, that Caney's account is a cosmopolitan one. But Page, whose own account of climate justice focuses on the rights and duties of states, criticises Caney, in part because it is not clear to Page (2008: 570) how Caney's 'methodological individualism' can be 'operationalised given the national focus of the current global climate architecture, or in the face of widespread belief in the political and ethical sovereignty of individual countries'. Page (2008: 570) asks, for example, how Caney can help us to know 'which individuals in the developed world should contribute'. Thus in Page's critique we again see two problems with extant doctrine of climate justice identified in previous chapters: the obsession with sovereign states and the focus on people in developed countries only.

How can we address these two preoccupations? As implied in the preceding chapter, Caney's brand of cosmopolitanism can offer some answers. We need not abandon the state, but we ought not fully accept the prevailing system either. Doing so in practice means muddling along while the Earth grows hotter and climate change does more harm and manifests itself in more human suffering. What is needed is an admission that the ethical sovereignty of states is not exclusive. A cosmopolitan corollary to the doctrine of international environmental justice could be a realistic and principled median between Page's (and other scholars') international climate justice, and some kind of global institution that, if ever created, will come about much too far in the future to help the world mitigate and cope with climate change in coming decades.

Using cosmopolitanism to help guide the world to better climate change policies requires us to focus much more attention on the world's polluters, including the new consumers. This alteration to the usual cosmopolitan response is perhaps radical, because cosmopolitans do not usually talk about duties in developing countries, but doing so is required by the rise of the new consumers. Furthermore, while practical cosmopolitans are right to recognise (if not endorse) the role of states, the orientation needs to change. Rather than advocating that states bear new burdens based on cosmopolitan morality – although they should do this – a more effective approach may be to benefit from cosmopolitanism's focus on the rights and duties of persons. States, rather than being the sole practical bearers or objects of cosmopolitan duties and rights, should instead be viewed more explicitly as facilitators of *individual* rights and duties. This approach may at first appear to be subtle, but its implications must be less so if the most disastrous consequences of climate change are to be avoided or significantly mitigated.

Putting cosmopolitanism into practice could mean a number of possible forms of governance and institutions. It could mean a world government, but it need not do so. As Hayden (2005: 21) points out, 'cosmopolitanism is not inherently opposed to the state *per se* . . . Rather cosmopolitanism is generally concerned to develop varied modes of governance – from the local to the global – with the goal of facilitating the rights and interests of individuals *qua* human beings. Indeed, states may be one mode of governance well suited to this end . . .'. David Schlosberg (2007: 188) argues that 'institutions of engagement' to bring about environmental justice 'could not exist solely at the state level; the focus must be at multiple levels – including both the state political realm and the transnational level'. While it might be ideal for the world to be governed by truly cosmopolitan institutions, they are unlikely any time soon. As Jon Mandle (2006: p. x) puts it, 'the world today and for the foreseeable future is one in which individuals and corporate actors pursue their goals against a background of rules and institutions created by states'. Axel Gosseries (2007: 280) believes that adopting the assumptions of cosmopolitanism need not stop us from 'using states as our point of reference [because states can be] most able to represent the individuals that constitute them and because they are currently the most relevant units in the context of global attempts of curbing [greenhouse gas] emissions'. Caney (2007) advocates a kind of 'revised statism' in which states do more to promote and implement global justice.

Future institutions could be premised on the understanding that the 'central argument of contemporary cosmopolitan political thought is that the demands of justice must be decoupled, at least to some degree, from the territorial bounds of the state' (Maltais 2008: 594). Thomas Pogge (2008: 192) argues that externalities like climate change 'bring into play the political human rights of . . . outsiders, thereby morally undermining the conventional insistence on an absolute right to national self-determination'. He proposes a dispersion of state sovereignty premised on cosmopolitan morality (Pogge 2008: 184). For Lorraine Elliott (2005: 497), 'an ecologically sensitive cosmopolitanism demands transnational environmental justice between people within a world society as well as, and possibly in preference to, international justice between states in an international society'. So we are left with something less than world government, and certainly a continuing role for states, but institutions informed by moral cosmopolitanism's advocacy of the equal worth of all persons and the need for protecting their basic rights. Consequently, following the moral cosmopolitan belief that 'every human being has a global stature as the ultimate unit of moral concern' (Pogge 2002c: 169), but without rejecting the state system generally and the climate change regime in particular, the cosmopolitan corollary to international doctrine seeks to establish a much more prominent place for human beings – their rights, responsibilities and duties – in the evolving climate change regime.

FEATURES OF THE COSMOPOLITAN COROLLARY

A cosmopolitan corollary to international doctrine would include several features. Andrew Linklater's description of a cosmopolitan, 'post-Westphalian' world is one based in part on the premise that 'the primary duty of protecting the vulnerable rests with the source of transnational harm and not with the national governments of the victims' (Linklater 1998: 84, quoted in Dobson 2005: 268–9). This implies that the source of the harm, and the responsibility not to cause harm in the first place, rests with actors other than national governments, or at least in addition to them. From a cosmopolitan perspective, this responsibility rests with people regardless of the national governments ruling the place where they live. Cosmopolitan diplomacy surrounding climate change should be premised on the rights and duties of human beings. Persons should be at the centre of climate change discourse, negotiations and policies, and they should be the viewed as the primary *ends* of diplomacy and government policy, comparable to Kant's imperative to treat human beings as ends rather than means to an end (that is, protecting state interests). Thus, while governments will inevitably retain a central role, as even most cosmopolitans recognise (at least because states are unlikely to surrender their role), unlike even most cosmopolitan arguments (which value persons but translate that into interstate duties, usually for rich states only), the cosmopolitan corollary would have governments playing the role of facilitators of global citizenship and the implementation of cosmopolitan obligation.

The ethical advantages are clear from a cosmopolitan perspective – human beings move to the centre. But equally important is the practical impact of more consumers and polluters being brought into the discourse of, and the solutions to, climate change. Very crucially, there are political benefits as well: states are, to some extent, taken off the hook, greatly reducing their political reflex to resist genuine collective action on climate change, in part because their citizens will at least see everyone in the world being treated more or less equally based on their conditions in life. Additional political advantages come from governments of the economic benefits that can accrue to many of their constituents from implementing the cosmopolitan corollary (see below).

The cosmopolitan corollary to the doctrine of international environmental justice starts with recognition of global justice and the rights of all persons and the duties of capable persons, and includes conscious efforts to actualise global justice in agreements and the national and multinational institutions of states. The corollary comprises two fundamental changes to the manner in which climate change is dealt with at the diplomatic level: (1) a change in the official discourse so that it acknowledges and affirms the rights and duties of all people in the context of climate change, and (2) the incorporation of human rights and responsibilities into international agreements on climate change. From this new perspective, people become primary objects and explicit ends of political and diplomatic negotiations, agreements and policies. International agreements and resulting policies would aim to promote a global environment suitable for human well-being and flourishing. States become facilitators of human rights and protectors of human interests related to climate change – a 'responsibility to protect' people and their 'human sovereignty' (to invoke two important concepts increasingly accepted by the international community) from the adverse impacts of climate change, in keeping with the notion that states exist to protect people and their well-being.

The cosmopolitan corollary mirrors Hayden's suggestion (2005: 132) that 'the discourse of sustainable *development* should give way to a discourse of sustainable or environmental *justice*'. The corollary is intended to be layered with extant national and international responses to climate change. That is, a global justice approach to addressing climate change is one that integrates discourse and thinking about, and action by, people as an add-on or supplement to more traditional communitarian approaches. As a corollary to the doctrine of international environmental justice, it is aimed at expanding the scope of climate justice by combining international justice with responses premised on world ethics. In short, human beings become a central moral basis of climate change diplomacy and policy, in effect making the climate change regime more ethically sound and more likely to elicit necessary action.

Elliott (2006: 363) rightly suggests that we must doubt whether 'relying on states not just as moral agents but as the moral subjects . . . is sufficient to ensure that individuals and their communities will be treated justly'. One might respond that we ought to do away with the role of states. But, even if that were a long-term aim, in the meantime a first step ought to be to promote individuals as moral agents even as states remain largely in control. This could be the starting point for an alternative to the state-centric tragedy of the atmospheric commons. The cosmopolitan corollary is, in essence, cosmopolitanism grafted onto extant Westphalian norms and the doctrines of international relations that have so far guided climate diplomacy and policy responses. In this respect it is a kind of bridge across the divide between the nation-state system and the imperative of climate protection. This process of making *both* persons (and their rights and responsibilities) and states (and their rights and responsibilities) objectives of the climate change regime is the central feature of the cosmopolitan corollary. By more explicitly encompassing both states and persons, it builds on and corrects the existing state-centred regime.

At first it may seem that the cosmopolitan corollary is nothing more than doing what many cosmopolitans already advocate: moral cosmopolitanism implemented by state institutions, whether they be at the national or the international level. But there is a fundamental point that bears emphasis: rather than being about how states can, at best, implement cosmopolitan principles, the cosmopolitan corollary is about how states can help *persons* implement cosmopolitan duties and enjoy related rights. As such, the cosmopolitan corollary is about moral ethics and practical ethics for people, while recognising that states cannot realistically be removed from the mix. This apparently subtle difference between what most cosmopolitans do (because they are more realistic than their critics assume) – accepting, perhaps grudgingly, states and their institutions – and what the cosmopolitan corollary is intended to do – to be closer to the cosmopolitan ideal of *individuals* being at the centre of policy or, put another way, more fully to view states as incidental to world ethics – is more important than it might seem. This is because the corollary has potential *practical and political* significance. If cosmopolitanism is used only to find new rights and duties for states, which it is used to do most often in the context of climate change, then it does little to help us escape the curse of Westphalia – the blame game among states of 'you go first' that has plagued the climate change negotiations – and might even make things worse by *reinforcing* the obligations of certain states over others, or at the very least affirming those obligations in the minds of the people that matter, such as government leaders and diplomats. Alternatively, the cosmopolitan corollary should have the effect of helping states to free themselves from the myopia of states' rights and obligations by focusing on how practical progress on climate change can be made through recognising and trying to actualise the rights and obligations of *affluent, capable persons everywhere*.

The cosmopolitan corollary to the doctrine of international environmental justice is principled, practical and politically viable – probably even politically essential. Most of the ethical-normative arguments for cosmopolitanism and world ethics can be brought to bear to justify bringing global justice into the climate change debate (see Chapter 5).

The cosmopolitan corollary to the doctrine of international environmental justice is therefore principled; it is the right thing to do for all the reasons outlined in previous chapters (for example, causality, capability, vital interests, harm and so forth). Perhaps most simply, the cosmopolitan corollary is more principled and just because it attaches duties to those who cause global warming and advances the rights of those who suffer the most from it, regardless of their nationality. The cosmopolitan corollary stops ignoring humans, their rights and their duties. Instead it aims to recognise and promote the rights of people, especially the least well-off, while incorporating the duties of all capable people, notably affluent ones all over the world, into efforts to address climate change. To be sure, for the cosmopolitan corollary to work – for it to contribute to ending the tragedy of the atmospheric commons that is the reality of a climate change regime premised on international doctrine – the principle of it will have to be taken seriously. This is not so much because principle matters, which of course it does for all sorts of reasons, but because if diplomats see cosmopolitanism as just another instrument for promoting state interests, they and their governments will fall back into the same tragic behaviour. By aiming the climate change regime at promoting cosmopolitan principles, which means putting people at the centre of the regime, diplomats can direct more attention to the causes and consequences of global warming and less to how traditional state-centric policies present problems for their states' perceived interests and long-held positions in the international relations of climate change.

The cosmopolitan corollary to the doctrine of international environmental justice is practical. It reflects climate change realities rather than assuming that the problem, and all the solutions to it, must or even can comport with the Westphalian assumptions of state sovereignty, rights, autonomy and independence. Unlike current doctrine, it focuses on the actual source of much of the world's greenhouse gas pollution – individuals. It directly addresses the increasingly important role played by large numbers of newly affluent people in the developing world while still fully encompassing affluent people in the industrialised world and while recognising that many poorer people in the latter are relatively minor contributors to the problem or not really capable of taking on obligations related to climate change. Pogge (2005: 53) has argued that people in wealthy countries ought not to avoid taking responsibility for the harm they cause to people in poor countries simply because *some* of the latter 'will get away with murder or with enriching themselves by starving the poor . . . This sad fact neither permits us (affluent people in affluent countries) to join their ranks, nor forbids us to reduce such crimes as far as we can.' We must agree with Pogge *and* take his argument further: we can no longer ignore affluent people in poor countries because it is no longer *some* of them committing 'murder' but now many millions of them. Put another way, the duty of most people in developed countries to act is undiminished, but we must stop letting all of the new consumers, and even rich elites in developing countries, get away with murder as their counterparts in the West have done for so long. Even if we accept that cosmopolitanism, world ethics and global justice are idealistic in other spheres of human activity, in the case of climate change they are utterly practical and necessary. The degree to which the cosmopolitan corollary is practical depends in large measure on how it is conceived and implemented. But this is just as true of other ideas for addressing climate change.

The cosmopolitan corollary is *politically viable* and likely to be politically essential if the world is to salvage the climate change regime and move it towards much more robust outcomes, especially in terms of mitigation but also in terms of adaptation and even compensation. It is perhaps important to reiterate at this point that what is being proposed is not to replace the doctrine of international environmental justice, but rather to supplement it and build upon it: to formulate and to implement a corollary to the international environmental justice doctrine. Thus the corollary is unlikely to face the kind of opposition from people and governments that would be experienced by wholesale change, and perhaps it would be embraced. What is more, if the world's new consumers are brought into the climate change regime, as the corollary would require, people in rich countries will see affluent people in poor countries responding. This will make it far easier for governments of the developed countries to sell the climate change regime to their citizens. At the same time, governments of developing countries can agree to limit the greenhouse pollution of their affluent citizens without undermining long-standing demands for international justice. They need not do what they insist that they will not do – take on mandatory *national* commitments to cut emissions of greenhouse gases. They can tell their citizens that they have won the argument in this respect. But the result on the ground is that large numbers of people (albeit affluent minorities) in developing countries will start to limit their emissions even as majorities of people there (the poor) are not required to do so and instead are aided to improve their living standards. Put another way, pollution among some groups *within* developing states will decline even though pollution *of* those states is not required to do so.

Additionally, insofar as affluent people everywhere contribute to global funds for dealing with climate change, the net effect will be more aid for people in developing countries (see below). Everyone could be equally obligated to pay in based on a 'cosmopolitan formula' that takes into consideration needs, capabilities, affluence, responsibility (and perhaps enjoyment of the fruits of past emissions) and so forth. While some funds would go to the poor in affluent countries, the net result is more funds going to help people in developing countries, making a new, more cosmopolitan regime much easier to sell there. This can be achieved without poor countries having to take on any new (and unfair) international burdens, yet they (actually, only some of their citizens) will be seen by rich states (and their citizens) to be doing so by paying into a new climate fund and by limiting, if not always reducing, their greenhouse gas emissions.

The cosmopolitan corollary is *not* politically idealistic, least of all utopian. The beauty of this approach is in part its political palatability. It gives states and diplomats the political cover they need by allowing them to stick with their long-standing principles. States qua states take on few new obligations when agreeing to the cosmopolitan corollary. Diplomats from developing states can go home and say, in all truthfulness, that they have not compromised on their demands for international justice, and even that people of the developed countries have, through their governments, agreed to take on new commitments to cut greenhouse gas emissions. Diplomats from developed states can go home and claim, accurately, that people in the developing countries have finally agreed, via their own governments, to take on new commitments to cut their greenhouse gas pollution. This gives an important political concession to the developing countries – they get respected and even paid as demanded and as is right, including from the perspective of international justice – and it gives the developed countries what they want and that which is required – involvement of the developing world in emissions cuts. The psychological impact of this new paradigm on populations in Australia, Canada, the United States and other developed countries where people (and governments) have been waiting for developing countries to commit to greenhouse gas cuts could be very powerful. It gives governments the political insulation they need finally to do what they know, as states, that they ought to have been doing for some time.

IMPLEMENTING THE COROLLARY AMONG STATES

At the international level, the cosmopolitan corollary would involve changes in diplomatic discourse; changes in international agreements that explicitly invoke, recognise and incorporate cosmopolitan rights and duties; changes in how those agreements are implemented by international institutions so that individuals' rights and duties are actualised; new kinds of funding mechanisms to distribute aid; and, ideally, representation of people as well as governments in negotiations. Even though the act of climate diplomacy is not cosmopolitan per se, comprising as it does relations among sovereign states and their representatives, implementing cosmopolitan imperatives should be one of its primary aims – not its only aim, not an aim to replace current aims premised on the seemingly inevitable baggage of extant international norms, not a utopian aim – but a practical aim based on the realities of climate change, its causes and its consequences. If states will not get out of the way (which they will not, realistically), they ought to at least become mediators of cosmopolitan duties; states and their international organisations ought to enable global citizenship, at least in the context of climate change, alongside national citizenship.

One assumes that states would prefer that we do not talk about climate change in cosmopolitan terms. States prefer the locus of rights and (albeit less so) obligations to remain squarely with them. Cosmopolitan justice threatens state sovereignty. While developing-country governments certainly welcome greater obligation on the part of affluent individuals in affluent states to provide aid to the world's poor, on top of wealthy governments' existing ethical and legal obligations to do so, many of those governments are unlikely to welcome burdens being placed on their own affluent people because the associated individual rights are generally anathema to those governments (and perhaps because that would include burdens for the individual governors themselves). To actualise cosmopolitan climate justice would probably bring into some question the good thing that a few developing states have now: concessional aid and investment linked to climate change. China, for example, has already experienced a minor windfall of investment under the climate change convention's Clean Development Mechanism. In 2005, the mechanism brought an additional $250 million investment into China, but by 2008 the country was receiving about $2.5 billion in annual investments through it (WWF 2008). With over half of all Clean Development Mechanism credits going to China, *all* the rest of the developing world is left to share the remainder, meaning that those countries are not receiving the benefits that especially the poorest among them have been hoping would come from the mechanism. Any suggestion that affluent Chinese ought to be paying out for those kinds of projects, as a matter of global justice, would be difficult for the government to accept unless it is part of a new cosmopolitan bargain that is perceived to be fair. The cosmopolitan corollary is such a bargain.

A key to the corollary is that requirements of global justice, such as recognition of human rights, distributive equity and increasing the capabilities of the disadvantaged, ought to be incorporated into agreements. What is more, procedural fairness, whereby *people* and their interests are part of the process of negotiating the agreements, should be implemented. According to the cosmopolitan corollary, the poor and underrepresented ought not be merely means to an end; they ought to be involved in the dialogue (as cosmopolitans often argue) and the *ends* of the climate change regime, in addition to ends focusing on states and other actors. It is important not only to empower poor states in climate change negotiations, but also to empower poor people, and indeed all people, or at the very least to try to represent their concerns, interests and even their aspirations. As a general rule, people significantly affected by political decisions and institutions have a right to be represented in making those decisions and running those institutions (cf. Pogge 2002c: 184). Importantly, diplomats and others should 'not look for a perfect system of representation before acting on the already obvious imperfect and biased system we have, and to bring a form of presence to those regularly left out of the decision-making process' (Schlosberg 2007: 195–6). Procedural fairness could involve proxies that might include representatives of various peoples, non-governmental actors in some cases, and even existing governments that have been seated through fully democratic means. One example of involving people more in negotiations about climate change can be found in the Aarhus Convention on Access to Information, Public Participation in Decision-Making and Access to Justice in Environmental Matters of the European Region (see Sachs 2002: 52). The Aarhus Convention provides legal remedies to enable citizens to challenge governments' denial of information and to increase participation of the public in international forums (Ebbesson 2007: 692, 701).

FUNDING GLOBAL CLIMATE JUSTICE

One important example of how the cosmopolitan corollary might be implemented among states is new (or reformed) funding mechanisms. While funding is not the only method for implementing the cosmopolitan corollary, it is worth exploring here as an illustration of what the corollary might look like in practice. Additionally, funding is important because it is related to other aspects of dealing with climate change: funds can enable greenhouse gas emissions limitations, and they can be used to help people and communities adapt to climate change and also to compensate people harmed by it. The act of gathering funds can also influence behaviours (for example, taxes on fossil fuels can discourage their use). Coming up with substantial funds, for example, to assist those harmed by climate change, need not be at all onerous. By way of illustration, Pogge (2005: 52) points out that a 'global resource dividend' (that is, a global tax) of only 1 per cent of global product would raise several hundred billion dollars. Importantly, his global resource dividend and taxes on excessive energy use for climate adaptation and compensation are not a form of aid: 'It does not take away some of what belongs to the affluent. Rather, it modifies conventional property rights so as to give legal effect to an inalienable moral right of the poor' (Pogge 2005: 52).

Many cosmopolitans argue in favour of different kinds of global funds, with proceeds dispersed so that everyone has enough resources for a dignified basic existence, for the protection of human rights or for 'unconditional basic income', particularly when basic needs are denied as a consequence of global structures, such as the world economic order (Brock and Brighouse 2005: 8). Brian Barry (1995: 153) suggests that cosmopolitanism is 'best satisfied in a world in which rich people wherever they lived would be taxed for the benefit of poor people wherever they live', thereby reducing the role of sovereign states while allowing them a role for raising funds and allowing international organisations a role for distributing those funds.[1] In principle, cosmopolitanism would suggest that global distribution of income be derived from an income tax 'levied at the same rate on people with the same income, regardless of where they live', and those who receive the resulting funds 'should be poor people again regardless of their place of residence' (Barry 1999: 40). This could be the general model for a tax related to climate change premised on global justice. Among specific measures could be a carbon tax on greenhouse gas emissions, which Barry says would ideally be collected 'directly from the users or polluters', which is preferable to taxing states based on per-capita national incomes, because 'individual income acts as a proxy for resource use wherever the person with income lives' (Barry 1998: 155). More of the money should come from earmarked taxes on non-essential activities related to climate change. This would include, among other things, taxing international airline flights, luxury goods and other non-essential polluting activities and goods.[2] This would raise new money and restrain harmful activity. Here we see the affluent aiding and acting to address climate change, in congruence with aims of cosmopolitan justice.

Peter Barnes (2001) has proposed a 'sky trust' that would administer greenhouse gas emissions rights on behalf of the world's citizens and pay out dividends to everyone based on its income. This could work by charging oil and coal companies for emissions rights, which would raise the cost of carbon intensive energy and require people who use it to pay higher prices, while also generating fees for the trust. This would reduce demand and generate income that could be distributed to everyone, meaning that those who use the least of the polluting energy sources (that is, the world's poorest people) would benefit the most. The fact that the funds would be redistributed to everyone, including the poor who need the higher-priced fuels, 'would help to gain public recognition for the idea that, despite different emission levels, all citizens have an equal *per capita* right to the atmosphere' (Sachs and Santarius 2007: 190). Hillel Steiner (2005: 36) describes a global fund, to which each state has an equal per-capita claim, that is derived from aggregating the individual claims of people in that state. He says that 'each person – regardless of where on the globe he or she resides – is owed that equal amount', the payment of which could take the form of 'unconditional basic income' or an 'initial capital stake' or some other *equal per-capita* form (Steiner 2005: 36). Steiner (2005: 36) believes that such a fund 'would serve to establish a variety of benign incentive structures informing relations both within and among nations'.

Jouni Paavola (2005: 317) argues for institutionalising responsibility for greenhouse gas emissions through a uniform carbon tax implemented at the national level. He recommends a low tax-free quota, which would put the bulk of the tax burden on already developed states, but he also advocates extending this tax to other countries 'when they become significant per capita greenhouse gas emitters' (Paavola 2005: 317). This latter point raises the question of what happens if per-capita emissions in developing countries go over the allowable global per-capita level, as many will, even while those countries remain poor. The answer should be to go ahead and implement the tax, qualifying it for poor people who have no choice but to exceed the acceptable level, but of course making sure that it captures the truly affluent in those poor states. As Paavola (2005: 317) suggests, his tax proposal would create incentives for efficient energy choices, and the resulting 'combined compensation and assistance fund' would provide revenue to be used 'for compensating the impacts of climate and for assisting adaptation to climate change'. Indeed, if poor people and communities in *all* countries could draw on the fund – that is, including people in developed countries – it could provide important incentives for their governments – most importantly, developed-country governments – to take on their responsibilities, and would certainly mitigate the disincentives they have for not doing so up to now.

The United Nations could administer funding to limit climate change and aid those who suffer from it the most. Some or much of the money raised from climate-related taxes could be deposited into existing funds, such as the Global Environment Facility, the Special Climate Change Fund, the Least Developed Country Fund and the Kyoto Protocol Adaptation Fund. These funding mechanisms are hardly ideal, but they are increasingly like the kind of thing that is needed. There might also be a new fund, perhaps a Future Climate Fund, specifically designed to aid future generations, possibly funded primarily from a tax on fossil fuels used by affluent people everywhere, to help future generations cope with climate change caused by past, present and future greenhouse gas emissions. A new climate-fund scheme would have an important difference compared to existing schemes: contributions to it would be based on individuals' responsibilities, duties and capabilities, and pay-outs, while administered by international organisations and capable and willing state governments (or perhaps non-governmental organisations when states are unwilling or unable to administer the funds, as occurs with some development and disaster aid), would be based on individual responsibilities, duties and capabilities. This would mean, for example, that a New York City executive would pay into the fund, but so too would a Hong Kong executive. In aggregate, of course, all affluent Americans (that is, the United States and most of its citizens) would pay far more into the fund than they would receive, and all Chinese (qua China) would receive far more than China's affluent citizens pay into the fund. But rich Chinese would no longer be treated just like poor ones – and very poor Americans would no longer be treated as though they are rich. Individuals can and should also give money to non-governmental organisations doing credible work to alleviate the suffering of those affected by climate change now and in the future.

Some countries in Europe have started to levy taxes and auction pollution rights, with some of the resulting funds going to climate adaptation programmes in developing countries. Twenty per cent of funds collected from the European Union's Emissions Trading Scheme (about $2 billion per year by 2020) are to be allocated to projects related to climate change, including adaptation in developing countries (*The Economist* 2008). One feature of the 2007 Bali Roadmap was a protocol for collecting a levy on Clean Development Mechanism projects, the funds from which are to be contributed to the Kyoto Protocol's adaptation fund. In 2008 governments agreed that the levy would amount to 2 per cent of the value of carbon credits that developed countries derive from the mechanism's projects in developing countries, with predictions that resulting funds would be as much as $950 million by 2012 (*The Economist* 2008). Alas, the actual amount of funding from this and other sources 'is just a puff of smoke' compared to the many tens of billions per year (at least) that poor countries and their people will need to adapt to climate change (*The Economist* 2008).[3]

The formula for whom should pay into climate change funds, and the amount they should pay, could be based on equal per-capita shares as a starting point, with other considerations being factored in (for example, responsibility, capability). The same kind of formula could apply to cuts, limits or increases in greenhouse gas emissions by and for individuals. (The majority of the world's people ought to be allowed and even empowered to increase their emissions while the world transitions to a post-carbon energy system.) The funds resulting from the cosmopolitan corollary could pay for things like disaster relief, poverty alleviation, sustainable development, greenhouse gas mitigation, adaptation measures, technology transfers and the like – and even for compensation. The ultimate ends of this funding must be human beings, although one assumes that some of the money might go to programmes to improve countrywide economies that would benefit as many people as possible going forward. One aim of the fund should be to discourage growth in populations where people become very affluent until affluence and fossil-fuel use can be decoupled (which is likely to happen without much encouragement because the most developed places tend to have low population growth, or even population decline). Importantly, as people in poor countries become wealthier, most of the benefits that accrue to them from the climate fund would go down. The aim would be for them one day to become affluent enough to stop receiving aid and to start paying into the fund, freeing up more money for those most in need, creating a kind of virtuous and increasingly effective cycle of funding for climate-related objectives.

The incentives and disincentives resulting from collecting such funds, if based on climate-related criteria such as energy use (causal criteria) and poverty (consequential criteria), could be used to promote greenhouse gas mitigation and the redistribution of wealth across borders, mainly to the advantage of people in developing states – thus giving their governments incentives to participate – and to the least well-off people everywhere, including in developed states – thus also giving their governments some incentive to participate, or at least lowering opposition to doing so. While such funds would be *international*, they would be fundamentally structured on cosmopolitan principles – on per-capita bases in terms of fundraising and payouts. There would have to be incentives and schemes for states unwilling or unable to implement the corollary as manifested in new agreements, and money might have to be given to non-governmental actors or international financial institutions to disperse (or to be put into escrow for when those actors are allowed by governments to aid people). Given that this responsibility to aid is largely (but not wholly) based on responsibility for suffering, affluent individuals in affluent states might have more obligation to provide aid because they often benefit more from their own and others' past pollution. However, the responsibility of affluent people in less affluent countries to aid starts from the moment they live an affluent life, and of course increases the more they consume and pollute – in addition to inherent obligations on all affluent and capable people, regardless of how much they pollute, to aid the poor. Importantly, none of this absolves affluent governments from continuing and increasing the types of international transfers that obtain at present or are already envisioned in the context of the climate change regime.

Actualising these or other sorts of schemes for funding and otherwise acting on the cosmopolitan corollary among states would admittedly run up against many of practical obstacles, but Barry confronts this head on: 'unless the moral case is made, we can be sure nothing good will happen. The more the case is made, the better the chance' (Barry 1998: 156).

IMPLEMENTING THE COROLLARY WITHIN STATES

While affluent individuals have obligations to act and to aid – while this is a cosmopolitan obligation – states remain very important. Short of people acting and organising themselves to implement cosmopolitan climate justice, we cannot ignore states. There is too little time for that. Luigi Bonanate (1995: 118) has a point when he writes that 'the application of criteria of justice is aimed (materially and effectively) at individuals but also that, *without* the mediation of the state, no justice can be really achieved'. Robyn Eckersley (2004) has argued for development of 'transnational green states' that are both ecologically responsible and act as facilitators of ecological citizenship. While her argument is pro-state, it is for a certain kind of state that accepts responsibilities for what happens beyond its borders and that 'would be more conducive than either traditional liberal or civic republican states to considering trans-species, transboundary, and intergenerational values and interests' (Eckersley 2004: 189). Development of such states would be a step towards implementing the cosmopolitan corollary and its environmental objectives. However, while moving in this direction is a good long-term goal, it is unlikely to happen in the decade or so that we have to tackle climate change robustly. An alternative to state-centric approaches to climate change is to combine what is good (or unavoidable) about the state with new strategies that address what is bad about it – namely, the failure of groups of states to work collectively to address climate change effectively.

ENCOURAGING, ENABLING AND REQUIRING DOMESTIC ACTION

How might the cosmopolitan corollary be implemented within existing states? Taxation, regulation and infrastructure come to mind. A primary objective should be to find a process whereby states help people to act on their cosmopolitan obligations as global citizens. Governments should assist the actualisation of cosmopolitan climate justice by more strictly *regulating* the non-essential polluting activities of affluent residents. Along with taxes, this will deter harmful behaviours and spur development of technologies that allow people to do things that make them truly happy without harming other people in the future. The most obvious activity to regulate is the use of fossil fuel energy – for example, by banning all but the most fuel-efficient cars (and eventually all cars using fossil fuels) and by restricting fossil-fuel-intensive recreation. Insofar as these regulations and the global taxes suggested above adversely affect poor people, as restrictions on international leisure travel might hurt people in poor parts of the world dependent on tourism, affluent governments and people should step in with assistance. The needs of present generations should not be ignored for those of the future, but the present does not trump the future, and even some poor people may have to rely on different forms of income in the light of the consequences for climate change.

Taxes and regulations are each government's sticks to persuade or force affluent people to live in ways that are consistent with cosmopolitan climate change obligations. There should also be carrots, such as generous tax rebates, subsidies and payments to encourage more environmentally benign activities. At the very least, governments ought not to create economic and other structures – and infrastructures – that make it more difficult or nigh impossible for individuals to act on their cosmopolitan obligations. An example is China's repeat of the mistake made in the West, especially the United States, of building highways and encouraging a car culture at the expense of mass transit.[4] Indeed, in China, bicycle lanes are being *removed* from cities to make way for cars, not installed, as sometimes now happens in Europe and North America. Governments ought instead to do more to create economic and physical infrastructures that are consistent with cosmopolitan climate justice. This would include creating efficient, comfortable and affordable mass-transit systems while making the use of cars less attractive in the medium and long term (unless a new climate-friendly personal transport vehicle is developed), and building distributions systems for alternative energy (perhaps hydrogen). Governments could also build new 'virtual' infrastructures, such as educational systems, strict building codes and accounting measures premised on a 'green', much more environmentally benign, future.

This would require new economic assumptions that are not premised on endless material consumption and economic growth. As Sachs (2000: 24) points out, 'in a closed environmental space, the claim for justice cannot be reconciled any longer with the promise of material-intensive growth, at least not for the world's majority. For this reason, the quest for justice will need to be decoupled from the pursuit of development with a capital "D".' Governments ought to start refocusing their societies and economies, through new economic policies and education, towards emphasising happiness over consumption (see Carley and Spapens 1998: 134–67). This effort would entail emphasising qualitative betterment over economic growth in already developed areas, with savings that come from new lifestyle choices available for investment in environmentally beneficial economic practices (Wapner and Willoughby 2005: 86). Equity and environmental sustainability can be achieved if 'people can revalue those forms of wealth which cannot be bought with a credit card: the enjoyment of quality, friendship, beauty [and the cherishing of] well-being rather than well-having' (Sachs 2002: 37). For developing countries, the 'right' to develop and pollute the way that affluent countries have done is trumped by the obligation and prudential need not to do so, not least because much better alternatives are available.

Massive investment in new energy technologies is an imperative to any successful battle against climate change. Societies and people will not stop using energy. However, we cannot sit back and put our faith in technology, at least not for the current generation. Technologies that would allow us to continue our current ways while reducing greenhouse gas emissions by 80 per cent *or more* – the latter being what scientists suggest is required without delay if the world is to avoid climate catastrophe – cannot realistically be created and implemented in time. In James Garvey's words (2008: 111), 'it's a ludicrous risk, a bet that we can continue with our lives as they are in the hope that something unknown or untested might make everything all right in the end'. As Garvey (2008: 106) argues, 'probably what we have to do – in addition to enormous technological efforts . . . – is change our lives. Instead of finding technological solutions for our energy needs, we have to find ways of needing less energy'.

One of the most important things that governments can do to help foster voluntary just action is to make climate change a central and priority subject at all levels of education. Peter Singer argues for educating young people to achieve environmental justice: 'In a society like America, we should bring up our children to know that others are in much greater need, and to be aware of the possibility of helping them, if unnecessary spending is reduced. Our children should also learn to think critically about the forces that lead to high levels of consumption, and to be aware of the environmental costs of this way of living' (Singer 2004: 164). How could anyone disagree with this proposal (not least because it appears that Singer's audience is Americans)? At the risk of being provocative, one might say that only a narrow-minded (free-market-obsessed) person could think otherwise. But what if we are to say the same of people in China and India? I am not certain, but I think Singer would say his argument applies to them as well. At present, however, millions of children are being raised in affluent families in developing countries whose educational systems are most decidedly not advocating restraint on consumption.

As a first step, a climate curriculum ought to be implemented with great urgency in developed countries, and ideally in all countries with effective and sufficiently funded educational systems. If climate change were to be given priority, all people, but especially younger ones, would become intimately attuned to the need for action and to precisely what they can do. Those young people would in turn have some influence on older people who are more resistant to change. In countries where educational systems are already not meeting demand, which is the case especially in many poor countries, the priority of developed-country governments, international organisations and particularly climate-related funds (such as those described above) should be to bolster those needy systems to help them educate people about climate change. It seems almost trite to say that more education is needed; few would disagree. But this needs to be implemented with extreme urgency. A major advantage of doing this is that it is more likely to foster action short of mandatory regulation, making things much easier for governments.[5]

Governments already do most of these things to varying, if grossly inadequate, degrees. But we need impetus for new policies based on cosmopolitan obligations and the contract that governments have with citizens to assist them in living a good life and to help them fulfil their obligations to other people, obligations that in this case will almost invariably benefit fellow citizens as well as foreign nationals. This is a case where doing the right (cosmopolitan) thing is not only good for others far away but also good for oneself and one's compatriots.

One increasingly popular proposal for action on climate change involves 'contraction and convergence' (Meyer 2000), which calls for per-capita emissions of each state to be brought to a level that is equal with other states and that the atmosphere can withstand, in practice meaning that emissions in wealthy countries would come down to a safe level (contraction) and those in most developing countries would go up to that level (convergence). The notion of contraction and convergence is essentially based on egalitarianism (Heyward 2007: 526). The equal per-capita amount that the atmosphere can withstand is roughly 1.5 tonnes of carbon-dioxide equivalent per person per year, which compares with about 20 tonnes in the United States, on one end of the spectrum, and about 0.1 tonnes in Mali, on the other (Smith 2006: 97). What the cosmopolitan corollary would require is that this policy be implemented not only among states but *within* them as well. This would mean that, while most people in rich countries would lower their greenhouse gas emissions to the globally safe per-capita level, most people in poor countries would be allowed to increase their emissions to that level. A difference between this approach and international doctrine is, of course, that poor people in wealthy states would not bear an unfair burden. Conversely, while most people in poor and developing countries would be allowed to increase their greenhouse gas emissions to the globally safe level, a large minority of people – the affluent – in those same countries would be required to reduce them. The precise amounts set for people would reasonably and fairly depend on their circumstances. Some people are in no position to reduce their emissions, and some emissions over the safe per-capita limit might be allowed for certain people if there is no alternative. At the same time, it is reasonable and probably necessary to expect some people to reduce their emissions below the globally safe level. The candidates for this requirement will be those who have polluted far more than they should have done already and who have the means (financial, technological and so forth) to reduce their emissions below the globally safe level while still meeting their basic needs.

One of the advantages of the cosmopolitan corollary to international climate justice is that it helps to get around a major problem associated with international agreements: there is no actor able to enforce them. When one thinks about the changes on the ground that will be required to address climate change adequately – when one contemplates the millions of sources of greenhouse gases and the billions of people involved – it is clear that no realistically imaginable global actor could force sufficient compliance. Positive incentives will be critically important to success. States have to want to comply, at least all but the smallest ones. Global climate justice, if implemented as suggested (or in similar ways) offers incentives – especially substantial new aid from global funds for purposes important to states and their populations – to get more governments and more people on board. This, in turn, eliminates some of the disincentives, such as seeing economic competitors avoid action, that have prevented many states from participating in serious international efforts to limit greenhouse gas emissions.

Like their counterparts in the developed states, the more affluent people in developing countries – those living like many Westerners by driving cars and living in large houses in the suburbs – have an obligation to cut their (fossil-fuel) energy use and to aid those in their own countries and beyond who will suffer from climate change. To be sure, they would be most likely to aid their fellow citizens, which is most practicable, but their obligation to aid those in other countries still obtains. Efforts to adopt the Western consumer culture in developing countries are misguided. There is ample evidence that this will not make affluent people any happier (Crocker and Linden 1998), and there is a strong argument that more wealth – more economic growth in national terms – is terrible for the environment even when not accompanied by more individual consumption (because money not spent, unless stuffed in a mattress, is invested in harmful material production) (Wapner and Willoughby 2005: 77–89). Basic material needs must be met, but basic needs of the world's vast poor ought not be trumped to fulfil the limitless emotional desires of the world's affluent people. Instead of living more like Americans, affluent people in developing countries ought to upstage them by showing how living simpler, more environmentally benign lives can make them happier and can be more rewarding. This, of course, applies fully to most people in the developed world as well.

CONCLUSION

At least in the context of climate change, cosmopolitanism and the cosmopolitan corollary to the doctrine of international environmental justice are principled, practical and politically viable. This contrasts with the purely interstate approach, which is ethically deficient, ignoring as it does the rights and obligations of people, and which is very limited in its practicality because it is premised on narrow state interests and thus has resulted in little collective action to combat climate change seriously, in the process stifling politically innovative, robust and truly effective solutions to climate change. As Elliott (2005: 498) notes, 'the normative interests of the state remain evident in the dominance of sovereignty claims and national interest that are pursued at the expense of cosmopolitan values and at the expense of the environment. The state therefore remains ambiguous as [a] cosmopolitan moral agent.' Even as we look to states to assist in fostering global environmental justice, it is worrying that they have utterly failed to arrest global warming and respond to climate change effectively, let alone doing so in a way that promotes justice, whether among states or among people. Consequently, while there is too little time to replace states, meaning that we must aim to reform their practices through injections of cosmopolitanism, the fundamental truth is that most of the important solutions to climate change are in the hands of people. Their role is the main subject of the next chapter.

To suggest that persons ought to become much more of a subject of diplomacy and policy among states is not radical. Indeed, as Christian Reus-Smit (2004: 7) has pointed out, one important feature of today's international system is a 'progressive "cosmopolitanisation" of international law, the movement away from a legal system in which states are the sole legal subjects, and in which the domestic is tightly quarantined from the international, toward a transnational legal order that grants legal rights and agency to individuals and erodes the traditional boundary between inside and outside'. One assumes that one thing driving this evolution in international law is a growing sense that persons per se have moral standing. This shift offers a good basis for the cosmopolitan corollary to international doctrine generally and international environmental justice in particular.

Nigel Dower identifies a key question that ought to guide climate diplomacy and policy: 'a cosmopolitan view will require us to look very hard at policies with a view to answering the question: does this contribute to or avoid not impeding the overall global good *vis-à-vis* the environment?' (Dower 2007: 186). To quote Onora O'Neill (2000: 2), 'any theory of justice that wishes to be taken seriously must respect empirical findings'. The doctrine of *international* justice and its underlying normative foundations have failed these tests. The statist doctrine has failed to give sufficient moral and practical weight – and usually fails even to acknowledge – the empirical reality of the shifting balance of greenhouse gas emissions away from being mostly a collective act of people in developed countries, which might once have made the climate change regime's focus on states practically reasonable, to one in which the emissions of affluent people in developing countries are rapidly rising towards those of people in developed countries. Thus an approach to justice that respects reality is one that at least incorporates, but ideally embraces, cosmopolitan ethics because it is that source of justice – it is *global* justice – that captures the full moral, practical and political facts of climate change in a globalised world.

NOTES

1. To avoid the familiar problem of the rich in poor countries stealing the funds, the transfers might have to be made directly to individuals rather than via governments.
2. As always, the super rich will simply pay taxes on activities that are not regulated.
3. The United Nations Development Programme (2007: 194) gives a 'lower ballpark' figure of $86 billion annually for costs of adaptation to climate change in 2015, while Christian Aid places costs at $100 billion per year (Flam and Skjaerseth 2009: 110).
4. Mass-transit systems are indeed being built in China, but the same infatuation with the car experienced in the United States and elsewhere – more highways and more ring roads – is developing with a vengeance (see Gallagher 2006).
5. Shallcross and Robinson (2006) describe an educational scheme to promote both global citizenship and environmental justice. For a plan to implement environmental education that contributes to 'ecological citizenship', see Dobson (2003: 174–207).

CHAPTER 8

THE UNAVOIDABILITY OF GLOBAL JUSTICE

Part I described the practical and ethical challenge of climate change. Chapter 1 summarised how global warming and climate change will become growing problems in the future, with the adverse impacts being felt most severely by those countries and people least responsible for causing them. Chapter 2 briefly examined how ethics and justice have become important in world affairs. Drawing on several accounts of justice, it showed how climate change is a profound matter of international and indeed global justice – or, more appropriately, *in*justice. Part II looked at justice in the context of international environmental agreements and regimes. Chapter 3 described the evolution of international environmental justice, which has been applied by states in the context of climate change because international justice has been perceived to be necessary for garnering the widest possible collective international action to limit global warming and to deal with its effects. As summarised in Chapter 4, governments have generally agreed on the importance of overarching justice principles, notably common but differentiated responsibility. However, the implementation of international climate justice by states has fallen far short of what is required ethically and practically, failing to address climate change in a robust way. Indeed, greenhouse gas emissions and the pace of climate change are *increasing* markedly. Taken together, the chapters in Part II located the failure of the climate change regime largely in its preoccupation with the rights and duties of *sovereign states*, and the consequent tragedy of the atmospheric commons resulting from the short-sighted logic of perceived national interests.

With that tragedy in mind, I have argued that there is a crucial *practical* role for world ethics in the international response to climate change. An approach to the problem based on world ethics and global justice may be much more politically viable than current practice. This is a case where doing what is right is also what is likely to achieve agreements among states and other actors that will be implemented with the desired effects.

Starting Part III, Chapter 5 identified and described an alternative way of viewing global climate change: cosmopolitanism. To be sure, international justice remains important. But it is far too narrow an approach because, in its myopic focus on states, it fails to recognise the locus of climate change – namely, people. This is a mistake that cosmopolitanism is well suited to overcome. Cosmopolitan concern with human rights and duties comports with the realities of climate change, in particular the role of hundreds of millions of individual polluters everywhere who pollute because they want to, not because they need to. These polluters include most people in the developed countries but also the rapidly growing population of new consumers in developing countries. As we saw in Chapter 6, cosmopolitanism not only directs our attention to much more than states; it does something that international climate doctrine has failed to do so far: recognise the practical and ethical significance of the rights and duties of all individuals, including the duties of many millions of capable and increasingly affluent people living outside the developed states who are becoming major contributors to the problem. Drawing on both the reality of the state-centred climate change regime and the desirability (moral and practical) of a cosmopolitan alternative, Chapter 7 described some features of a cosmopolitan corollary to international environmental justice in the context of climate change. The corollary is an attempt to correct biases built into the climate change regime by bringing people into debates and policies at both national and international levels. If states are to do more to address climate change effectively, they ought to help facilitate cosmopolitan justice.

However, even if the cosmopolitan corollary is actualised by states, we cannot leave it all up to them. Something else is needed for global climate justice to be achieved soon: citizen action. Consequently, in this chapter I briefly return to those actors who are by definition central to cosmopolitanism and who should be, by necessity, central to action on climate change: human beings.[1]

GLOBAL CITIZENSHIP

The historical evolution of justice beyond borders has progressed through several stages that look something like this: first, states had very few if any obligations to other states, apart from non-intervention and respecting emissaries. This stage lasted well into the twentieth century. Second, affluent states accepted some obligations to poor states, certainly to aid in case of widespread famine and major natural disasters. This stage arguably became entrenched in the last half-century, in large part because technology and modern transport enabled states to aid one another relatively easily. Third, affluent states accepted (although they have too rarely acted upon) some obligations to individuals abroad who are very badly off, such as those persons suffering from endemic poverty or widespread human rights abuses. We now seem to be in a fourth stage, in which there is some agreement that affluent *individuals in wealthy states* have obligations to people in poor states suffering from severe poverty and other major ills. This is an important and positive development manifested in many governments' official development assistance and the work of non-governmental organisations. What we should hope for now is an extension of this to include the obligations of affluent individuals *everywhere* – that is, a fifth stage of *cosmopolitan* justice that does not merely see people in poor countries as objects of assistance but, if they are affluent, also sees them as objects of obligation to end, as much as possible, harm to others and to assist those who are badly off in their own countries *and* elsewhere. This fifth stage – truly global justice – may be vital if the world is to address climate change effectively.

Thomas Pogge (Pogge 2008: 209) summarises the injustices of climate change: 'The global poor get to share the burdens resulting from the degradation of our natural environment while having to watch help-lessly as the affluent distribute the planet's abundant natural wealth amongst themselves.' This points to an important conclusion: if we are to address global climate change successfully, we will have to acknowl-edge that justice extends well beyond borders. Our future requires that our responses to the globalisation of environmental changes and their consequences include a globalisation of justice. This need not mean that global justice replace justice within states, but it does mean that justice beyond them can no longer be given minor consideration when solutions to climate change are deliberated, formulated and implemented. Global justice – the rights of all people everywhere to their due, and the duties of people everywhere depending on their capabilities – will have to be at the centre of all aspects of climate change politics and policy.

Robin Attfield (2005: 42) argues that, if cosmopolitan ethics are required, so too is 'an awareness of global citizenship, capable of motivating matching action and a corresponding sense of identity, rather as national citizenship has often served to motivate national patriotism and corresponding forms of communitarianism'. Global citizenship, which involves commitment to a global ethic that transcends national borders, 'is morally as important as it has always been, and for practical purposes increasingly urgent too' (Attfield 2003: 160). The notion that people might have and act on a sense of global citizenship is not new. Hayden (2005: 15) notes that even the Stoics rejected 'the mutual exclusion of the local and the world communities, and thought it both possible and desirable for individuals to consider themselves citizens of their local communities as well as citizens of the world'.

The idea of world citizenship emphasises the 'individual's duty to act with consideration for the environment, for distant strangers and for unborn generations [and] defends the ideal that there are obligations to avoid acting in ways that result in the domination and exploitation of other peoples' (Linklater 2002: 264). Andrew Dobson (2004: 1) argues that 'ecological citizenship and global justice are intimately linked [and] when properly understood they entail each other'. Good global citizens ought to 'accept that one's actions may have indirect (and largely unintended) effects on both distant and future peoples', with the significance of those actions lying 'partly in their contribution to cumulative impacts' (Dower 2003: 93). Angel Saiz (2006: 13) believes that notions of global citizenship emerge from 'green thought', because environmental issues challenge traditional conceptions of citizenship, requiring as they often do transnational responses. Janna Thompson (2001: 145) speaks of the planetary citizen: 'someone who assumes her share of responsibility for the collective achievement of good which she and virtually everyone else values'. According to Derek Heater (1996: 215), 'in the case of global environmental citizenship, the right of access to and enjoyment of a common planetary environment is matched by the obligations of conservation and a whittling down of national sovereignty . . . Without the obligation on the present generation to conserve, the global environmental rights of future generations will be infringed.'

While a sense of global citizenship is probably on the rise, it is not likely to be strong enough or widespread enough, or to come soon enough, to lead to the kind of responses that are necessary to avert the worst effects of climate change. Consequently, states have their role in fostering individual behaviours in keeping with the duties of global citizenship. This is necessarily a contrived global citizenship, but difficult times such as this require that we live with contradictions of modern life rather than ignore them at our peril, and indeed at the peril of others weaker and more vulnerable than ourselves. Cosmopolitanism and the cosmopolitan corollary to international doctrine ought to be squarely focused on what defines the underlying moral theory: the universal rights *and* responsibilities of persons. Ideally, of course, people would voluntarily act on their duties as global citizens. To start with, affluent people who pollute more than necessary (which means almost all of us who are affluent) would reduce their environmental footprints. But this is a tall order for the period of time in the near future that we have left to begin very seriously tackling climate change. Consequently, states will have to be part of the process whereby people act as global citizens even while a sense of duty as global citizenship evolves, as it almost certainly must if the future is to be bearable for the worlds' poor and vulnerable.

Thompson (2001) acknowledges that finding the right ways to act on planetary responsibilities will not be easy, and that states will still be in the picture. But, even if the means by which individuals could fully realise 'their role as planetary citizens' do not exist, they can still 'aim toward this idea and try to make it a reality' (Thompson 2001: 144). 'Planetary citizenship' and the cooperation it engenders provide 'at least a psychological and moral basis for transcending' differences that may arise from, for example, national affiliations (Thompson 2001: 145). Indeed, as Attfield (2005: 47) points out, the common reliance of all people on the global environment 'has the implication that the different nations, creeds and communities are bound together by shared interests, awareness of which can increase people's motivation to recognise their global citizenship'. Climate change does not give us time to wait for a culture of planetary or world citizenship to develop slowly, but the cosmopolitan corollary at least offers states incentives, or at least mitigates and removes many political obstacles, to start developing the necessary institutions, to lay down new codes of conduct and to encourage their citizens to 'think globally, act locally'.

INDIVIDUAL RESPONSIBILITY

From the cosmopolitan perspective, individual persons have funda-mental rights that precede 'rights' of states. This will come as a comfort to those whose fundamental rights, such as the right to subsistence, are violated as a consequence of climate change. But cosmopolitanism also identifies persons as *moral agents* with duties to act in certain ways (Evans 2003: 25). If this agency is to effect a reduction in global warming and the resulting injustices of climate change, affluent people every-where will have to live differently. For example, we will have to enjoy airline travel much less, or not at all, because it quickly puts us over our fair share of lifetime greenhouse gas emissions. One easy new behaviour that the affluent could adopt would be to stop eating animals, because meat production uses large amounts of fossil-fuel energy and produces methane, a potent greenhouse gas. Affluent individuals also ought to push for political and economic changes that will lead to widespread environmental action by more individuals (Wapner and Willoughby 2005). Even where this is not so easy, as in authoritarian states, affluent individuals can still act to restrain their consumption, thereby contributing to what should be a global collective effort of the affluent. Put another way, just because one cannot change one's national system or because the international climate change regime does not yet encourage or enable individual responsibility, there is no excuse to live like most Americans or Australians. To do so, at least in a material sense, is immoral; it is a violation of cosmopolitan justice and a recipe for climate disaster.

At present, many of us follow (although we seldom admit as much) the moral concept of 'us-here-now': 'to deny that we have obligations to any but the present generation or those living now, to deny that we have obligations to non-humans, and to deny we have obligations to human beings outside our own society' (Dower 1998: 161).[2] This is unjust. Living a life of *sufficiency* is the better ethical and environmental course. While affluent global citizens need not completely renounce their material way of life, we ought to 'weaken our relentless pursuit of and attachment to it. So we need to re-evaluate our commitment to material affluence, for the sake of the environment, for the sake of peace and for the sake of the poor' (Dower 2007: 210). Wolfgang Sachs (2001) identifies an obligation of the affluent everywhere: 'the global middle class, which includes Southern elites, have got to search for forms of well-being which are capable of justice'. He argues that

> the move toward models of frugal use of wealth among the affluent is a matter of equity, not just of ecology. However, conventional development thinking implicitly defines equity as a problem of the poor. But [in] designing strategies for the poor, developmentalists [have] worked towards lifting the bottom – rather than lowering the top. The wealthy and their way of producing and consuming weren't under scrutiny, and the burden of change was solely heaped upon the poor. In future, however, justice will be much more about changing lifestyles of the rich than about changing those of the poor. (Sachs 2000: 25)

We ought to consume what we need from the Earth to survive and to fulfil our basic needs, and perhaps a bit more, doing all that we reasonably can to limit the impact of that consumption. The affluent ought to consume what we need, full stop. By behaving this way, affluent individuals *everywhere* would be actualising global justice and acting as good global citizens.

Affluent individuals also ought to *aid* those who are suffering from climate change and those who will suffer in the future. We should, along with governments, aid current sufferers because we have probably benefited from economic wealth that was generated from past environmental exploitation and that is causing present harm. We should aid people who will suffer in the future because our emissions of greenhouse gases will harm them, particularly the poor. As Thomas Pogge (1998: 510) puts it, 'those, usually the affluent, who make more extensive use of the resources of our planet should compensate those who, involuntarily, use very little' because the 'better-off – we – are *harming* the worse-off insofar as the radical inequality we uphold excludes the global poor from a proportional share of the world's natural resources and any equivalent substitute' (Pogge 2005: 40).[3]

Cosmopolitan climate justice means that obligations to act on climate change, and to aid those (individuals) harmed by it, apply to nearly all affluent individuals regardless of where they live. If governments do more by way of using taxes, regulations, infrastructure and education to change behaviours, many people will be pushed to do the right thing. However, if governments are not fully up to the task (which could be the case until environmental conditions grow very bad indeed), affluent individuals will have to find it within themselves to act on cosmopolitan obligations. Insofar as possible given where we live and the structures that rule our lives, we should act responsibly by cutting our greenhouse gas emissions if we are already emitting more than our fair share (as we almost certainly are) or, if we are not emitting much more than our fair share of greenhouse gases, by limiting them to somewhere near that level. Even if it is not clear where this limit should be set, affluent people should do everything we reasonably can to limit our greenhouse pollution. Non-essential polluting activities should be avoided. The increasingly common practice of paying for 'carbon offsets', usually by giving money to non-governmental organisations that support forests and other carbon sinks, is not an adequate response for capable global citizens because, as Hermann Ott and Sachs (2002: 173) put it, 'the cosmopolitan notion focuses on self-limitation for the sake of a good global neighbourhood'. According to Brian Orend (2006: 212), 'being a good international [and, we might add, *world*] citizen *does demand some self-restraint*'.

In sum, if a person's emissions of greenhouse gases are above an acceptable global per-capita average (currently and for their lifetime) *and* his basic needs are met and he is significantly above the poverty level in his local community (how this level is defined is, of course, important), then he has an obligation to bring those emissions at least down to (or as near as possible to, given his circumstances) an acceptable global per-capita amount that would prevent climate upset. If his personal lifetime emissions exceed his share of the global per-capita limit, he is also obliged to aid those who suffer from climate change now and especially in the future (at least insofar as his excess emissions were something he could control), and arguably he ought to do this even if his personal emissions have been low. Obligation to act – to limit our own contributions to greenhouse gas concentrations in the atmosphere – is a negative responsibility because we 'participate in, and profit from, the unjust and coercive imposition' (Pogge 1998: 502) of climate change on those who will suffer from it the most, and the obligation to *aid* those who are suffering or will suffer from climate change is a positive responsibility because we 'could improve the circumstances' (Pogge 1998: 502) of those sufferers (to give only two justifications among a number, as we have seen).

We may be products of our local communities, but this fact need not preclude us from having communal sentiments and even strong loyalties to more than one community. We can have strong loyalties to family members *and* to a larger community (for example, neighbourhood, town, village, nation, state). We can also have loyalties towards both our national community and the global community. Indeed, some cosmopolitans have argued that we fall within a number of concentric circles, each associated with a different kind of loyalty, attachment or identify – the self and the family being closest, humanity being farthest away (but still there). While we may be more likely to care about and assist fellow family members or fellow citizens of our national community, it does not follow that we would be willing to do so – or ought to do so morally – regardless of the costs to others far away (Brown 1992: 186). Today we are increasingly the products of international society; many young people especially are immersed in, shaped by, and identify with ideas and ideals, personalities, styles and forms of popular culture that have no boundaries. Insofar as these things shape people's identities, they can foster greater sympathy for, and affiliation with, the larger world community. What is more, a sense of local community loyalty may be expanded to a sense of global community when focused by menaces that remind us that we all live in one world. If Martians were to attack Earth and collective action were required to repel that invasion and to cope with it, I suspect that the sense of community among all humanity would be much stronger. Climate change is analogous to an attack from Mars, albeit in slow motion.

Additionally, sympathies to a larger community – a sense of altruism towards people in other countries, especially the poor – can be a by-product of specific types of domestic communities (Lumsdaine 1993). Countries that have generous arrangements for domestic social welfare (for example, the Scandinavian states) are especially generous with their official development assistance. Not coincidentally, these are the same countries that have been most forthcoming with efforts to help less affluent countries develop in an environmentally sustainable fashion. Countries that are generous at home are generous abroad. This shows that we may be a product of the domestic communities in which we were raised, but this does not mean that we will not have a feeling of obligation to other people far away. One's domestic community can even cultivate such sentiments. This is an important development, suggesting, perhaps paradoxically, that (some) national communities themselves may be in front of (most) persons in seeing that a globalised world cannot be governed effectively if we fail to acknowledge and act upon, in a universal way insofar as practicable, the moral worth of every human being.

If one is not moved by the need to act for the well-being of others, it is worth being reminded that there is almost nothing to lose and much to gain from doing what global environmental justice demands. Despite dramatic increases in average income and gross domestic product per person in the developed countries of the world over the last half-century, people's satisfaction with life and happiness has not increased, demonstrating that affluence, as opposed to meeting one's real needs, is not directly linked to people's feelings of well-being (Speth 2008: 129–34). Contrary to common wisdom, once people's basic needs have been met, plus a cushion for security and some modest luxuries, money and consumption do not buy happiness. But helping others does have its rewards in the form of self-satisfaction, and consuming only what we need gives us the knowledge that the environment upon which everyone relies, including ourselves, is more likely to be able to sustain us.

As James Garvey (2008: 150) points out, when one enters a funk from the apparent hopelessness of doing anything oneself to fight climate change, it is worth bearing in mind that the effects of one's behaviour are measured over a lifetime:

> against the claim that individual choices cannot matter much, is that nothing else about you stands a chance of making a moral difference at all. If anything matters, it's all those little choices. This rejoinder shows up all over the place, just about anywhere you hear the claim that nothing a single person can do could possibly make a difference. The little effects are the only effects you'll ever have. The only chance you have of making a moral difference consists in the individual choices you make.

The total impact of a life lived high on the hog compared to one lived simply adds up, and, when multiplied by a billion or more other relatively affluent people in the world, the impact is gargantuan. It is the difference between a livable planet for all and truly monumental suffering for billions.

CONCLUSION

As every day passes and increasing numbers of people join the ranks of the world's affluent classes, a cosmopolitan ethic of climate change becomes more urgent. *International* justice is necessary but inadequate, and the focus on it by diplomats, activists and scholars may be part of the problem. This points to the need for ways of transgressing sovereignty and for solutions that do not depend upon the state-centric assumptions that prevail in climate change negotiations and policies. The preoccupation with states is, of course, understandable. We live in a world of states in which national governments try to do what is required to manage common problems. But this preoccupation is a kind of myopia or even a psychosis of sorts; it is a mental straitjacket that 'forces' us to think of problems and their solutions in terms of states instead of in terms of human-based causes, consequences and remedies.

Without ignoring the role of states and other institutions to actualise climate justice, we should stop talking almost exclusively about national responsibilities and obligations. We should talk much more about individual obligations (of affluent persons) and consider these obligations when making policies and attempting to educate people about climate change. We also ought to spread the burden and stop letting affluent people in certain places avoid responsibility. Most people in the affluent countries are, of course, the most to blame. But it may be counterproductive to keep telling the Western middle classes that they should drive their cars less while they watch the developing world's roads fill with the same vehicles.

Even those who call for incorporating equity and justice into solutions to climate change usually do so in terms of states: justice among states is required because climate change is an injustice among them. This emergence of *international* environmental justice is an important step forward historically. Alas, it is only a small step because it has been actualised only minimally owing to the selfishness of its objects – states themselves – and because it ignores so many people who cause and experience the injustices of climate change. Even after more than two decades of international negotiations and much earnestness on the part of diplomats, government officials and activists, there is very little to show for all this activity relative to the scale of climate change and its severe impacts. Very little has come from the climate change regime so far by way of concrete and deep cuts in greenhouse gas emissions required to stem global warming, let alone the very large transfers of funds and technology to poor countries and people needed to spread the cuts widely and to help the most vulnerable cope with inevitable climate change. By ignoring individuals and indeed global justice, the climate change regime has backed states into a corner. This was all predictable given the nature of states to promote their narrow, usually short-term interests most of the time – at times even over the interests of their own citizens. Thus we need to go beyond international justice to consider fully, and to implement fully, *global* justice as well. We are all in this together, which implies quite a lot for every capable person, along with capable states, international organisations and other actors.

In this book I have not advocated what many cosmopolitans aspire to – world government. But it may come to that. Without alternatives to failed climate change policies based upon international doctrine, we may very soon reach a point where a global supranational entity is the only way to overcome the tragedy of the atmospheric commons. Even David Miller (2007: 269), hardly a proponent of cosmopolitanism, has said that, 'if global warming accelerates to the point where the continuance of human life in anything like its present form becomes doubtful, people might be willing to sign a Hobbesian global contract giving a central authority the power to impose fierce environmental controls on all societies'. But before it comes to that, 'we need to ask what might motivate ordinary people to impose the necessary restrictions on themselves' (D. Miller 2007: 269). One thing that might motivate them, as well as their governments, is a new contract that places cosmopolitan aims, and persons everywhere, at the centre of the climate change regime.

The general question of international climate justice is settled. Nobody is arguing very vigorously that the developed countries do not have special obligations. Much debate about the details remains, to be sure. However, the general question of global, *cosmopolitan* climate justice is still very much unsettled. We have not decided whether certain *people* have responsibility for justice towards others, especially if those people and the others we are concerned about are both living in poor countries. At the very least, insofar as one accepts a simple standard of ethics that identifies behaviour harmful to others in this context as being wrong, we have, by definition, an ethical deficit. We have devoted so much diplomatic and philosophical capital to arguing for international justice that we have avoided looking at the actual locus of environmental harm, which is largely the individual and, from an ethical perspective, especially the affluent individual with a major impact on climate and a choice about whether to end or exacerbate that impact. The solution to our ethical deficit, and to climate change, is, at least in large part, cosmopolitan justice. Ultimately what that means is a combination of political *and* personal morality, and behaviour to match.

We might not be able to solve climate change – we might not be able to prevent most of the adverse effects, including the monumental human suffering and economic costs it will entail in coming decades – if we do not embrace *global* justice. Doing so may be ethically, practically and politically *essential*. Insofar as climate change is the most important problem facing humankind, as well as other species sharing this planet, there can be no more profound argument in favour of world ethics. Human survival and world ethics go hand in hand; it is unlikely that we can have the former without the latter.

NOTES

1. We might accomplish the same environmental objectives even more successfully if we were to put all species at the centre of climate policy, but one assumes that bringing in non-humans to this extent would not have the political advantages of the human-centred cosmopolitan corollary. Thus doing so might not be the most realistic approach, given what little time we seem to have left to avert catastrophe.
2. I am using Dower's wording here, not his sentiments. As a cosmopolitan, he rejects this conception.
3. The radical inequality is defined as the better-off enjoying significant advantages of, and the worse-off being excluded from and not being compensated for a lack of access to, the natural resource base (see Pogge 2002c, 2008).

REFERENCES

Almond, Brenda (1995), 'Rights and justice in the environment debate', in David E. Cooper and Joy A. Palmer (eds), *Just Environments*, London: Routledge.

Anderson, Kevin, and Alice Bows (2008), 'Reframing the climate change challenge in light of post-2000 emission trends', *Philosophical Transactions of the Royal Society A* (doi: 10.1098/rsta.2008.0138), http://www.tyndall.ac.uk/publications/journal_papers/fulltext.pdf: 1–20.

Annan, Kofi (2006), 'Citing "frightening lack of leadership" on climate change', http://www.un .org/News/Press/docs/2006/sgsm10739.doc.htm.

Arthur, John, and William H. Shaw (1978), *Justice and Economic Distribution*, Englewood Cliffs, NJ: Prentice Hall.

Associated Press (2008), 'Global warming pollution increases by 3 percent', *International Herald Tribune*, 25 September, http://www.iht.com/articles/ap/2008/09/25/america/ Warming-Emissions.php.

Attfield, Robin (1999), *The Ethics of the Global Environment*, Edinburgh: Edinburgh University Press.

Attfield, Robin (2003), *Environmental Ethics*, Cambridge: Polity Press.

Attfield, Robin (2005), 'Environmental values, nationalism, global citizenship and the common heritage of humanity', in Jouni Paavola and Ian Lowe (eds), *Environmental Values in a Globalising World*, London: Routledge, pp. 75–104.

Baer, Peter (2002), 'Equity, greenhouse gas emissions, and global common resources', in Stephen H. Schneider, Armin Rosencranz and John O. Niles (eds), *Climate Change Policy*, Washington, DC: Island Press, pp. 393–410.

Barkdull, John, and Paul G. Harris (1998), 'The land ethic: A new philosophy for international relations', *Ethics and International Affairs*, 12: 159–78.

Barnes, Peter (2001), *Who Owns the Sky?*, Washington, DC: Island Press.

Barry, Brian (1989), *Theories of Justice*, Berkeley and Los Angeles: University of California Press.

Barry, Brian (1995), *Justice as Impartiality*, Oxford: Clarendon Press.

Barry, Brian (1998), 'International society from a cosmopolitan perspective', *International Society*, Princeton: Princeton University Press, pp. 144–63.

Barry, John (1999), 'Statism and nationalism: A cosmopolitan critique', in Ian Shapiro and Lea Brilmayer (eds), *Global Justice*, New York: New York University Press, pp. 12–66.

Barry, John (2008), 'Foreword', in Steve Vanderheiden (ed.), *Political Theory and Global Climate Change*, Cambridge, MA: MIT Press, pp. vii–x.

Beckman, Ludvig, and Edward A. Page (2008), 'Perspectives on justice, democracy and global climate change', *Environmental Politics*, 17/4: 527–35.

Beitz, Charles (1979a), 'Global egalitarianism: Can we make out a case?' *Dissent*, 26/1: 59–68.

Beitz, Charles (1979b), *Political Theory and International Relations*, Princeton: Princeton University Press.

Beitz, Charles (1983), 'Cosmopolitan ideals and national sentiment', *Journal of Philosophy*, 80: 591–600.

Beitz, Charles (1991), 'Sovereignty and morality in international affairs', in David Held (ed.), *Political Theory Today*, Stanford, CA: Stanford University Press, pp. 236–54.

Beitz, Charles (1999), 'International liberalism and distributive justice: a survey of recent thought', *World Politics*, 51/2: 269–96.

Benedick, Richard Elliott (1998), *Ozone Diplomacy*, Cambridge, MA: Harvard University Press.

Bentham, Jeremy (1962), 'Principles of international law', in John Bowring (ed.), *The Works of Jeremy Bentham*, vol. 2, New York: Russell and Russell, pp. 535–60.

Biermann, Frank (1996), ' "Common Concern of Humankind": The emergence of a new concept of international environmental law', *Archiv des Volkerrechts*, 34/4: 426–81.

Biermann, Frank (2005), 'Between the United States and the South', Global Governance Working Paper No. 17, Amsterdam: Global Governance Project.

Bonanate, Luigi (1995), *Ethics and International Politics*, Columbia, SC: University of South Carolina Press.

Boyle, Alan E. (1994), 'The Convention on Biological Diversity', in Luigi Campiglio et al. (eds), *The Environment after Rio*, London: Graham and Trotman, pp. 111–30.

Brahic, Catherine, and Reuters (2008), 'China warns of huge rise in emissions', *New Scientist Environment*, 22 October, http://www.newscientist.com/article/dn15011.

Brock, Gillian (2007), 'Global distributive justice, entitlement and desert', in Daniel Weinstock (ed.), *Global Justice, Global Institutions*, Calgary: University of Calgary Press, pp. 109–38.

Brock, Gillian (2009), *Global Justice*, Oxford: Oxford University Press.

Brock, Gillian, and Harry Brighouse (2005), 'Introduction', in Gillian Brock and Harry Brighouse (eds), *The Political Philosophy of Cosmopolitanism*, Cambridge: Cambridge University Press, pp. 1–9.

Brock, Gillian, and Darrel Moellendorf (2005), 'Introduction', *Journal of Ethics*, 9: 1–9.

Brown, Chris (1992), *International Relations Theory*, New York: Columbia University Press.

Bullard, Robert D. (1990), *Dumping in Dixie*, Boulder, CO: Westview Press.

Burhenne, Wolfgang E. (1992), 'Biodiversity: Legal aspects', *Environmental Policy and Law*, 22/5–6: 324–6.

Caldwell, Lynton K. (1990), *International Environmental Policy*, Durham, NC: Duke University Press.

Caney, Simon (2005a), 'Cosmopolitan justice, responsibility, and global climate change', *Leiden Journal of International Law*, 18: 747–75.

Caney, Simon (2005b), *Justice beyond Borders*, Oxford: Oxford University Press.

Caney, Simon (2006a), 'Global distributive justice and the environment', in Ronald Tinnevelt and Gert Verschraegen (eds), *Between Cosmopolitan Ideals and State Sovereignty*, London: Palgrave Macmillan, pp. 51–63.

Caney, Simon (2006b), 'Cosmopolitan justice, rights and global climate change', *Canadian Journal of Law and Jurisprudence*, 19/2: 255–78 [Lexis Nexis Academic: paras 1–57].

Caney, Simon (2007), 'Cosmopolitanism, democracy and distributive justice', in Daniel Weinstock (ed.), *Global Justice, Global Institutions*, Calgary: University of Calgary Press, pp. 29–64.

Caney, Simon (2008), 'Human rights, climate change, and discounting', *Environmental Politics*, 17/4 (August): 536–55.

Carley, Michael, and Philippe Spapens (1998), *Sharing the World*, New York: St. Martin's Press.

Central Intelligence Agency (2005), 'Field listing: GDP per capita', *The World Factbook*, http://www.cia.gov/cia/publications/factbook/fields/2004.html.

Chatterjee, Pratap, and Mathias Finger (1994), *The Earth Brokers*, London: Routledge.

China Daily (2004), 'China to have 140 million cars by 2020', *China Daily*, 4 September, http://www.chinadaily.com.cn/ english/doc/2004–09/04/content_371641.htm.

Climate Institute (2007), *Evidence of Accelerated Climate Change*, Sydney: Climate Institute.

Cochran, Molly (1999), *Normative Theory in International Relations*, Cambridge: Cambridge University Press.

Crocker, David A., and Toby Linden (1998) (eds), *Ethics of Consumption*, New York: Rowman and Littlefield.

Cronin, Ciaran, and Pablo De Greiff (2002), 'Introduction: Normative responses to current challenges of global governance', in Pablo De Greiff and Ciaran Cronin (eds), *Global Justice and Transnational Politics*, Cambridge, MA: MIT Press, pp. 151–95.

Cullen, Bernard (1992), 'Philosophical theories of justice', in Klaus R. Sherer (ed.), *Justice: Interdisciplinary Perspectives*, Cambridge: Cambridge University Press, pp. 15–64.

Dobson, Andrew (1998), *Justice and the Environment*, Oxford: Oxford University Press.

Dobson, Andrew (2003), *Citizenship and the Environment*, Oxford: Oxford University Press.

Dobson, Andrew (2004), 'Ecological citizenship and global justice: Two paths converging?' in Anne K. Haugestad and J. D. Wulfhorst (eds), *Future as Fairness*, Amsterdam: Rodopi, pp. 1–15.

Dobson, Andrew (2005), 'Globalisation, cosmopolitanism and the environment', *International Relations*, 19/3: 259–73.

Dobson, Andrew (2006), 'Thick cosmopolitanism', *Political Studies*, 54: 165–84.

Dower, Nigel (1992), 'Sustainability and the right to development', in Robin Attfield and Barry Wilkins (eds), *International Justice and the Third World*, London: Routledge, pp. 93–116.

Dower, Nigel (1997), 'World ethics', in Ruth Chadwick, *Encyclopedia of Applied Ethics*, San Diego: Academic Press, pp. 561–70.

Dower, Nigel (1998), *World Ethics*, Edinburgh: Edinburgh University Press.

Dower, Nigel (2000), 'Global ethics', in Adrian Hastings (ed.), *The Oxford Companion to Christian Thought*, Oxford: Oxford University Press, pp. 265–8.

Dower, Nigel (2003), *An Introduction to Global Citizenship*, Edinburgh: Edinburgh University Press.

Dower, Nigel (2007), *World Ethics*, 2nd edn, Edinburgh: Edinburgh University Press.

Dupont, Alan (2008), 'The strategic implications of climate change', *Survival*, 50/3: 29–54.

Dyer, Gwynne (2006), 'Adding fuel to the fire', *South China Morning Post*, 21 April: A15.

Earth Negotiations Bulletin (1997), 'Report of the Third Conference of the Parties of the Framework Convention on Climate Change: 1–11 December 1997', *Earth Negotiations Bulletin*, 12/76.

Ebbesson, Jonas (2007), 'Public participation', in Daniel Bodansky, Jutta Brunnee and Ellen Hey (eds), *The Oxford Handbook of International Environmental Law*, Oxford: Oxford University Press, pp. 681–703.

Eckersley, Robyn (2004), *The Green State*, Cambridge, MA: MIT Press.

The Economist (2005), 'Cars in China: Dream machines', *The Economist*, 2 June, http://www.economist.com/business/displaystory.cfm?story_id= 4032842.

The Economist (2008), 'Climate change and the poor: Adapt or die', *The Economist*, 11 September, http://www.economist.com/world/international/displaystory.cfm?story_ id = 12208005&fsrc = rss.

Elliott, Lorraine (2005), 'Transnational environmental harm, inequity and the cosmopolitan response', in Peter Dauverge (ed.), *Handbook of Global Environmental Politics*, Cheltenham: Edward Elgar, pp. 486–501.

Elliott, Lorraine (2006), 'Cosmopolitan environmental harm conventions', *Global Society*, 20/3 (July): 345–63.

Ellis, Anthony (1992), 'Utilitarianism and international ethics', in Terry Nardin and David R. Mapel (eds), *Traditions of International Ethics*, Cambridge: Cambridge University Press, pp. 158–79.

Evans, Mark (2003), 'World citizenship and the ethics of individual responsibility', *International Journal of Politics and Ethics*, 3/1: 25–43.

Fabre, Cecile (2007), 'Global distributive justice: An egalitarian perspective', in Daniel Weinstock (ed.), *Global Justice, Global Institutions*, Calgary: University of Calgary Press, pp. 139–64.

FAO (2007), Food and Agricultural Organisation, *Livestock's Long Shadow*, Rome: Food and Agriculture Organisation, http://www.fao.org/docrep/010/a0701e/a0701e00.htm.

Flam, Karrine Haegstad, and Jon Birger Skjaerseth (2009), 'Does adequate financing exist for adaptation in developing countries?', *Climate Policy*, 9: 109–14.

Forst, Rainer (2001), 'Towards a critical theory of transnational justice', *Metaphilosophy*, 32/1–2: 160–79.

Gallagher, Kelly Sims (2006), *China Shifts Gears*, Cambridge, MA: MIT Press.

Gardiner, Stephen M. (2001), 'The real tragedy of the commons', *Philosophy and Public Affairs*, 30/4: 387–416.

Gardiner, Stephen M. (2004), 'Ethics and global climate change', *Ethics*, 114/3: 555–600.

Garner, Jonathan (2006), *The Rise of the Chinese Consumer*, Chichester: John Wiley and Sons.

Garvey, James (2008), *The Ethics of Climate Change*, London: Continuum.

Gasper, Des (2005), 'Beyond the international relations framework: An essay in descriptive global ethics', *Journal of Global Ethics*, 1/1: 5–23.

Global Carbon Project (2008), 'Carbon budget and trends 2007', 26 September, http://www. globalcarbonproject.org/carbontrends/index.htm.

Gore, Al (2006), *An Inconvenient Truth*, Emmaus, PA: Rodale.

Gosseries, Axel (2007), 'Cosmopolitan luck egalitarianism and the greenhouse effect', in Daniel Weinstock (ed.), *Global Justice, Global Institutions*, Calgary: University of Calgary Press, pp. 279–310.

Hansen, James, et al. (2008), 'Target atmospheric CO_2: Where should humanity aim?' *Open Atmospheric Science Journal*, http://arxiv.org/pdf/0804.1126v2.

Hardin, Garrett (1968), 'The tragedy of the commons', *Science*, 162: 1243–8.

Harris, Paul G. (1996), 'Considerations of equity and international environmental institutions', *Environmental Politics*, 5/2 (Summer): 274–301.

Harris, Paul G. (1997a), 'Affluence, poverty and ecology: Obligation, international relations and sustainable development', *Ethics & the Environment*, 2/2 (Fall): 121–38.

Harris, Paul G. (1997b), 'Environment, history and international justice', *Journal of International Studies*, 40: 1–33.

Harris, Paul G. (1999a), 'Common but differentiated responsibility: The Kyoto Protocol and United States policy', *New York University Environmental Law Journal*, 7/1: 27–48.

Harris, Paul G. (1999b), 'Environmental security and international equity: Burdens of America and other great powers', *Pacifica Review*, 11/1: 25–42.

Harris, Paul G. (2000a) (ed.), *Climate Change and American Foreign Policy*, London: Palgrave Macmillan.

Harris, Paul G. (2000b), 'Defining international distributive justice: Environmental considerations', *International Relations*, 15/2 (August): 51–66.

Harris, Paul G. (2001a), *International Equity and Global Environmental Politics*, Aldershot: Ashgate.

Harris, Paul G. (2001b) (ed.), *The Environment, International Relations, and US Foreign Policy*, Washington, DC: Georgetown University Press.

Harris, Paul G. (2002a), 'Global warming in Asia-Pacific: Environmental change vs international justice', *Asia-Pacific Review*, 9/2 (November): 130–49.

Harris, Paul G. (2002b) (ed.), *International Environmental Cooperation*, Boulder, CO: University Press of Colorado.

Harris, Paul G. (2002c), 'Sharing the burdens of environmental change: Comparing EU and US policies', *Journal of Environment and Development*, 11/4 (December): 380–401.

Harris, Paul G. (2003a), 'Fairness, responsibility, and climate change', *Ethics and International Affairs*, 17/1: 149–56.

Harris, Paul G. (2003b) (ed.), *Global Warming and East Asia*, London: Routledge.

Harris, Paul G. (2004), ' "Getting rich is glorious": Environmental values in the People's Republic of China', *Environmental Values*, 13/2: 145–65.

Harris, Paul G. (2006), 'The European Union and environmental change: Sharing the burdens of global warming', *Colorado Journal of International Environmental Law and Policy*, 17/2: 309–55.

Harris, Paul G. (2007a), 'Collective action on climate change: The logic of regime failure', *Natural Resources Journal*, 47/1: 195–224.

Harris, Paul G. (2007b) (ed.), *Europe and Global Climate Change*, Cheltenham: Edward Elgar.

Harris, Paul G. (2008a), 'Climate change and global citizenship', *Law & Policy*, 30/4 (October): 481–501.

Harris, Paul G. (2008b), 'Climate change and the impotence of international environmental law: Seeking a cosmopolitan cure', *Penn State Environmental Law Review*, 16/2 (Winter): 323–68.

Harris, Paul G. (2008c), 'Implementing climate justice', *Journal of Global Ethics*, 4/2 (August): 121–40.

Harris, Paul G. (2008d), 'Introduction: The glacial politics of climate change', *Cambridge Review of International Affairs*, 21/4 (December): 455–64.

Harris, Paul G. (2009) (ed.), *The Politics of Climate Change*, London: Routledge.

Hayden, Patrick (2005), *Cosmopolitan Global Politics*, Aldershot: Ashgate.

Heater, Derek (1996), *World Citizenship and Government*, London: Macmillan.

Held, David (1995), *Democracy and Global Order*, Cambridge: Polity Press.

Held, David (2000), 'Regulating globalization? The reinvention of politics', *International Sociology*, 15 (June): 394–408.

Held, David (2005), 'Principles of cosmopolitan order', in Gillian Brock and Harry Brighouse (eds), *The Political Philosophy of Cosmopolitanism*, Cambridge: Cambridge University Press, pp. 10–27.

Heyward, Madeleine (2007), 'Equity and international climate change negotiations: A matter of perspective', *Climate Policy*, 7: 518–34.

Hoffman, Stanley (1981), *Duties beyond Borders*, Syracuse, NY: Syracuse University Press.

Hourcade, Jean-Charles, and Michael Grubb (2000), 'Economic dimensions of the Kyoto Protocol', in Joyeeta Gupta and Michael Grubb (eds), *Climate Change and European Leadership*, London: Kluwer Academic Publishers, pp. 173–202.

Human Rights Council (2008), 'Report of the Human Rights Council on its seventh session', New York: United Nations, www2.ohchr.org/english/bodies/hrcouncil/docs/ 7session/A-HRC-7-78.doc.

International Energy Agency (2006), *World Energy Outlook 2006*, Paris: International Energy Agency.

IPCC (2002), Intergovernmental Panel on Climate Change, *Climate Change 2001*, Cambridge: Cambridge University Press.

IPCC (2007a), Intergovernmental Panel on Climate Change, *Climate Change 2007: Synthesis Report*, http://www.ipcc.ch/ipccreports/ar4–syr.htm.

IPCC (2007b), Intergovernmental Panel on Climate Change, 'Summary for policymakers', in IPCC, *Climate Change 2007:Impacts, Adaptation and Vulnerability*, Cambridge: Cambridge University Press, pp. 7–22.

IPCC (2007c), Intergovernmental Panel on Climate Change, 'Summary for policymakers', in IPCC, *Climate Change 2007: Mitigation*, Cambridge: Cambridge University Press, pp. 1–23.

IPCC (2007d), Intergovernmental Panel on Climate Change, 'Summary for policymakers', in IPCC, *Climate Change 2007: The Physical Science Basis*, Cambridge: Cambridge University Press, pp. 1–18.

Jagers, Sverker C., and Goran Duus-Otterstrom (2008), 'Dual climate change responsibility: On moral divergences between mitigation and adaption', *Environmental Politics*, 17/4: 576–91.

Jamieson, Dale (1997), 'Global responsibilities: Ethics, public health, and global environmental change', *Indiana Journal of Global Legal Studies*, 5/1: 99–119.

Jamieson, Dale (2002), *Morality's Progress*, Oxford: Clarendon Press.

Jamieson, Dale (2005), 'Duties to the distant: Aid, assistance, and intervention in the developing world', *Journal of Ethics*, 9: 151–70.

Jones, Charles (1999), *Global Justice*, Oxford: Oxford University Press.

Kant, Immanuel ([1785] 1948), *The Moral Law*, trans. Herbert J. Paton, London: Hutchinson.

Karunakaran, C. E. (2002), 'Clouds over global warming', *CorpWatch*, 24 October, http://www.corpwatch.org/article.php?id = 4548.

Krasner, Stephen D. (1985), *Structural Conflict*, Berkeley and Los Angeles: University of California Press.

Larson, David L. (1994), *Security Issues and the Law of the Sea*, Lanham, MD: University Press of America.

Leopold, Aldo ([1949] 1968), *A Sand County Almanac and Sketches Here and There*, Oxford: Oxford University Press.

Lichtenberg, Judith (1981), 'National boundaries and moral boundaries: A cosmopolitan view', in Pater Brown and Henry Shue (eds), *Boundaries*, Lanham, MD: Rowman and Littlefield, pp. 79–100.

Linklater, Andrew (1998), *The Transformation of Political Community*, Cambridge: Polity.

Linklater, Andrew (1999), 'The evolving spheres of international justice', *International Affairs*, 75/3: 473–82.

Linklater, Andrew (2001), 'Citizenship, humanity and cosmopolitan harm conventions', *International Political Science Review*, 22/3: 261–77.

Linklater, Andrew (2002), 'Cosmopolitan harm conventions', in Steven Vertovec and Robin Cohen (eds), *Conceiving Cosmopolitanism*, Oxford: Oxford University Press, pp. 254–67.

Linklater, Andrew (2006), 'Cosmopolitanism', in Andrew Dobson and Robyn Eckersley (eds), *Political Theory and the Ecological Challenge*, Cambridge: Cambridge University Press, pp. 109–30.

Luard, Evan (1994), *The United Nations*, New York: St. Martin's Press.

Lumsdaine, David H. (1993), *Moral Vision in International Politics*, Princeton: Princeton University Press.

Lutken, Soren Ender, and Axel Michaelowa (2008), *Corporate Strategies and the Clean Development Mechanism*, Cheltenham: Edward Elgar.

McGrew, Anthony (2008), 'Globalization and global politics', in John Baylis, Steve Smith and Patricia Owens (eds), *The Globalization of World Politics*, 4th edn, Oxford: Oxford University Press, pp. 14–33.

Mahathir Mohamad (1998), 'Statement to the UN Conference on Environment and Development', excerpted in Ken Conca and Geoffrey D. Dabelko (eds), *Green Planet Blues*, 2nd edition, Boulder, CO: Westview Press, pp. 325–6.

McKibben, Bill (1989), *The End of Nature*, London: Anchor Books.

Maltais, Aaron (2008) 'Global warming and the cosmopolitan political conception of justice', *Environmental Politics*, 17/4: 592–609.

Mandle, Jon (2006), *Global Justice*, Cambridge: Polity.

Marburger, John (2008), 'A global framework: International aspects of climate change', *Harvard International Review* (Summer): 48–51.

Mason, Michael (2005), *The New Accountability*, London: Earthscan.

Mathews, H. Damon, and Ken Caldeira (2008), 'Stabilizing climate requires near-zero emissions', *Geophysical Research Letters*, vol. 35 (doi: 10.1029/2007GL032388).

Matsui, Yoshiro (2002), 'Some aspects of the principle of "Common but differentiated responsibility"', *International Environmental Agreements*, 2/2: 151–71.

Meyer, Aubrey (2000), *Contraction and Convergence*, Bristol: Green Books.

Midgley, Mary (2001), 'Individualism and the concept of Gaia', in Ken Booth, Tim Dunne and Michael Cox (eds), *How Might We Live*, Cambridge: Cambridge University Press, pp. 29–44.

Miller, David (1976), *Social Justice*, Oxford: Clarendon Press.

Miller, David (2007), *National Responsibility and Global Justice*, Oxford: Oxford University Press.

Miller, Marian A. L. (1995a), 'The Third World agenda in environmental politics', in Manochehr Dorraj (ed.), *The Changing Political Economy of the Third World*, Boulder, CO: Lynne Rienner.

Miller, Marian A. L. (1995b), *The Third World in Global Environmental Politics*, Boulder, CO: Lynne Rienner.

Moellendorf, Darrel (2002), *Cosmopolitan Justice*, Boulder, CO: Westview Press.

Mott, Richard (1993), 'The GEF and the Conventions on Climate Change and Biological Diversity', *International Environmental Affairs*, 5/4: 299–312.

Muller, Benito (2002), *Equity in Climate Change*, Oxford: Oxford Institute for Energy Studies.

Myers, Norman, and Jennifer Kent (2003), 'New consumers: The influence of affluence on the environment, *Proceedings of the National Academy of Sciences of the United States*, 100/8: 4963–8.

Myers, Norman, and Jennifer Kent (2004), *The New Consumers*, London: Island Press.

Netherlands Environmental Assessment Agency (2008), 'China contributing two thirds to increase in CO_2 emissions', 13 June, http://www.mnp.nl/en/service/pressreleases/2008/20080613Chinacontributingtwothirdsto increaseinCO2emissions.html.

New Scientist (2006), 'Climate change is all around us', *New Scientist*, No. 2543, http://environment.newscientist.com/channel/earth/mg18925432.600-editorial-climate-change-is-all-around-us.html.

Nordhaus, William (2008), *A Question of Balance*, London: Yale University Press.

Okereke, Chukwumerije (2008), *Global Justice and Neoliberal Environmental Governance*, London: Routledge.

O'Neill, Onora (1986), *Faces of Hunger*, London: Allen and Unwin.

O'Neill, Onora (1988), 'Hunger, needs, and rights', in Stephen Luper-Foy (ed.), *Problems of International Justice*, Boulder, CO: Westview Press, pp. 67–83.

O'Neill, Onora (1993), 'Kantian ethics', in Peter Singer (ed.), *A Companion to Ethics*, Oxford: Blackwell, pp.175–85.

O'Neill, Onora (1996), 'Ending world hunger', in William Aiken and Hugh LaFollette (eds.), *World Hunger and Morality*, Upper Saddle River, NJ: Prentice Hall, pp. 85–112.

O'Neill, Onora (2000), *Bounds of Justice*, Cambridge: Cambridge University Press.

Orend, Brian (2006), 'Good international citizenship', in Ronald Tinnevelt and Gert Verschraegen (eds), *Between Cosmopolitan Ideals and State Sovereignty*, London: Palgrave Macmillan, pp. 209–20.

Ott, Hermann E., and Wolfgang Sachs (2002), 'The ethics of international emissions trading', in Luiz Pinguelli-Rosa and Mohan Munasinghe (eds), *Ethics, Equity and International Negotiations on Climate Change*, London: Edward Elgar, pp. 159–77.

Ott, Hermann E., Bernd Brouns, Wolfgang Sterk and Bettina Wittneben (2005), 'It takes two to tango: Climate policy at COP 10 in Buenos Aires and beyond', *Journal for European Environmental and Planning Law*, 2: 84–91.

Paavola, Jouni (2005), 'Seeking justice: International environmental governance and climate change', *Globalizations*, 2/3: 309–22.

Page, Edward A. (2006), *Climate Change, Justice and Future Generations*, Cheltenham: Edward Elgar.

Page, Edward A. (2008), 'Distributing the burdens of climate change', *Environmental Politics*, 17/4: 556–75.

Park, Jacob (2005), 'Global climate change: Policy challenges, policy responses', in Dennis Pirages and Ken Cousins (eds), *From Resource Scarcity to Ecological Security*, Cambridge, MA: MIT Press, pp. 165–86.

Parks, Bradley C., and J. Timmons Roberts (2006), 'Environmental and ecological justice', in Michelle M. Betsill, Kathryn Hochstetler and Dimitris Stevis (eds), *Palgrave Advances in International Environmental Politics*, Houndmills: Palgrave Macmillan, pp. 329–60.

Paterson, Matthew (1994), 'International justice and global warming', paper for the Conference on Ethics and Global Change, Reading University, 29 October.

Paterson, Matthew (1996), 'International justice and global warming', in Barry Holden (ed.), *The Ethical Dimensions of Global Change*, New York: St. Martin's Press, pp. 181–201.

Paterson, Matthew (1997), 'Equity or justice', in Detlef Sprinz and Urs Luterbacher (eds), *International Relations and Global Climate Change*, Potsdam: Potsdam Institute for Climate Impact Research.

Pellow, David Naguib, and Robert J. Brulle (2005) (eds), *Power, Justice, and the Environment*, Cambridge, MA: MIT Press.

Perkins, Richard (2008), 'Incentivizing climate mitigation: Engaging developing countries', *Harvard International Review* (Summer): 42–7.

Pew Centre on Global Climate Change (2007), 'Summary of COP13', http://www.pewclimate.org/docUploads/Pew%20Center_COP%2013%20Summary.pdf.

Pogge, Thomas (1992), 'Cosmopolitanism and sovereignty', *Ethics*, 103: 48–75.

Pogge, Thomas W. (1998), 'A global resources dividend', in David A. Crocker and Toby Linden (eds.), *Ethics of Consumption*, New York: Rowman and Littlefield, pp. 501–36.

Pogge, Thomas (2002a), 'Cosmopolitanism: A defence', *Critical Review of International Social and Political Philosophy*, 5/ 3: 86–91 [proof].

Pogge, Thomas (2002b), 'Human rights and human responsibilities', in Pablo De Greiff and Ciaran Cronin (eds), *Global Justice and Transnational Politics*, Cambridge, MA: MIT Press, pp. 151–95.

Pogge, Thomas (2002c), *World Poverty and Human Rights*, Cambridge: Polity.

Pogge, Thomas (2005), 'Real world justice', *Journal of Ethics*, 9: 29–53.

Pogge, Thomas W. (2008), *World Poverty and Human Rights*, 2nd edn, Cambridge: Polity.

Rafferty, Kevin (2008), 'Jobs today or a planet tomorrow?' *South China Morning Post*, 18 October: A13.

Raupach, Michael R., et al. (2007), 'Global and regional drivers of accelerating CO_2 emissions', *Proceedings of the National Academy of Sciences of the United States of America*, 104/24: 10288–93.

Rawls, John (1971), *A Theory of Justice*, Cambridge, MA: Harvard University Press.

Rawls, John (1999), *The Law of Peoples*, Cambridge, MA: Harvard University Press.

Rawls, John ([1993] 2008), 'The law of peoples', in Thomas Pogge and Darrel Moellendorf (eds), *Global Justice*, St Paul, MN: Paragon House, pp. 421–60.

Reus-Smit, Christian (2004), 'Introduction', in Christian Reus-Smit (ed.), *The Politics of International Law*, Cambridge: Cambridge University Press, pp. 1–13.

Roberts, J. Timmons, Kara Starr, Thomas Jones and Dinah Abdel-Fattah (2008), 'The reality of official development assistance', Oxford: Oxford Institute for Energy Studies.

Russett, Bruce, and Harvey Starr (2004), 'International actors: States and other players on the world stage', in Daniel J. Kaufman et al. (eds), *Understanding International Relations*, Boston: McGraw Hill, pp. 43–60.

Sachs, Wolfgang (2000), 'Development: The rise and decline of an ideal', Wuppertal Paper No. 108, http://www.wupperinst.org/Publikationen/WP/WP108.pdf.

Sachs, Wolfgang (2001), 'EcoEquity interview with Wolfgang Sachs', http://www.ecoequity. org/ceo/ceo_3_4.htm.

Sachs, Wolfgang (2002) (ed.), *The Jo'burg Memo*, Berlin: Heinrich Boll Foundation.

Sachs, Wolfgang, and Tilman Santarius (2007), *Fair Future*, London: Zed.

Saiz, Angel Valencia (2006), 'Globalisation, cosmopolitanism and ecological citizenship', in Andrew Dobson and Angel Valencia Saiz (eds), *Citizenship, Environment, Economy*, London: Routledge, pp. 7–22.

Satz, Debra (2005), 'What do we owe the global poor?' *Ethics and International Affairs*, 19/ 1: 47–54.

Sands, Philippe (1994), 'The "greening" of international law: Emerging principles and rules', *Global Legal Studies Journal*, 1/ 2: 293–323.

Schlosberg, David (2007), *Defining Environmental Justice*, Oxford: Oxford University Press.

Shallcross, Tony, and John Robinson (2006), 'Education for sustainable development as applied global citizenship and environmental justice', in Tony Shallcross and John Robinson (eds), *Global Citizenship and Environmental Justice*, Amsterdam: Rodopi, pp. 175–93.

Shue, Henry (1992), 'The unavoidability of justice', in Andrew Hurrell and Benedict Kingsbury (eds), *The International Politics of the Environment*, Oxford: Oxford University Press, pp. 373–97.

Shue, Henry (1993), 'Subsistence emissions and luxury emissions', *Law and Policy*, 15 /1: 39–59.

Shue, Henry (1995), 'Equity in an international agreement on climate change', in Richard Samson Odingo, Alexander L. Alusa, Fridah Mugo, Joseph Kagia Njihia and Anne Heidenreich (eds), *Equity and Social Considerations Related to Climate Change*, Nairobi: ICIPE Science Press, pp. 385–92.

Shue, Henry (1996a), *Basic Rights*, 2nd edn, Princeton: Princeton University Press.

Shue, Henry (1996b), 'Solidarity among strangers and the right to food', in William Aiken and Hugh LaFollette (eds), *World Hunger and Morality*, Upper Saddle River, NJ: Prentice Hall, pp. 113–32.

Shue, Henry (1999), 'Global environment and international equity', *International Affairs*, 75/3: 531–45.

Shukla, P. R. (1999), 'Justice, equity and efficiency in climate change: A developing country perspective', in Ferenc L. Toth (ed.), *Fair Weather?*, London: Earthscan, pp. 145–59.

Singer, Peter (1996), 'Famine, affluence and morality', in William Aiken and Hugh LaFollette (eds), *World Hunger and Morality*, Upper Saddle River, NJ: Prentice Hall, pp. 229–43.

Singer, Peter (2003), 'Books for breakfast', transcript of remarks to the Carnegie Council on Ethics and International Affairs, 29 October, http:// cceia.org/viewMedia.php/prmTemplate ID/8/prmID/164.

Singer, Peter (2004), *One World*, New Haven: Yale University Press.

Singer, Peter (2006), 'Ethics and climate change: A commentary on MacCracken, Toman and Gardiner', *Environmental Values*, 15/ 3: 415–22.

Smith, D. Mark (2006), *Just One Planet*, Rugby: Intermediate Technology Publications.

Smith, Mark J., and Piya Pangsapa (2008), *Environment and Citizenship*, London: Zed.

Speth, James Gustave (2008), *The Bridge at the Edge of the World*, London: Yale University Press.

Steiner, Hillel (2005), 'Territorial justice and global redistribution', in Gillian Brock and Harry Brighouse (eds.), *The Political Philosophy of Cosmopolitanism*, Cambridge: Cambridge University Press, pp. 28–38.

Stern, Nicholas (2007), *The Economics of Climate Change*, Cambridge: Cambridge University Press.

Stone, Christopher D. (2004), 'Common but differentiated responsibilities in international law', *American Journal of International Law*, 98/2: 276–301.

Thompson, Janna (1992), *Justice and World Order*, London: Routledge.

Thompson, Janna (2001), 'Planetary citizenship: the definition and defence of an ideal', in Brendan Gleeson and Nicholas Low (eds), *Governing for the Environment*, Houndmills: Palgrave, pp. 135–46.

Tilly, Charles (1975) (ed.), *The Formation of National States in Western Europe*. Princeton: Princeton University Press.

Tin, Tina (2008), 'Climate change: Faster, stronger, sooner', Brussels: WWF, http://assets. panda.org/downloads/wwf_science_paper_october_2008.pdf.

Tornblom, Kjell (1992), 'The social psychology of distributive justice', in Klaus R. Sherer (ed.), *Justice*, Cambridge: Cambridge University Press, pp. 177–236.

UNFCCC (1992), United Nations Framework Convention on Climate Change, http://unfccc.int/resource/docs/convkp/conveng.pdf.

United Nations (1972), *Declaration of the United Nations Conference on the Human Environment* (A/CONF.48/14), New York: United Nations.

United Nations (1981), *Declaration on the Inadmissibility of Intervention and Interference Intervention in the Internal Affairs of States* (A/RES/36/103), New York: United Nations.

United Nations (1982), *United Nations Convention on the Law of the Sea* (A/CONF.62/122 (1982), annex III, New York: United Nations.

United Nations (1983), *Law of the Sea*, New York: United Nations.

United Nations (1989), *United Nations Conference on Environment and Development*, General Assembly Resolution No. 44/228, New York: United Nations, http://www.un.org/ documents/ga/res/44/a44r228.htm.

United Nations (1992), 'Convention on Biological Diversity', Montreal: Secretariat of the Convention on Biological Diversity, http:// www.cbd.int/doc/legal/cbd-un-en.pdf.

United Nations (1993), 'Rio Declaration on Environment and Development', in United Nations, *Report of the United Nations Conference on Environment Development*, vol. 1, New York: United Nations Publications, pp. 3–8.

United Nations (1995), 'Berlin Mandate', New York: United Nations, http://unfccc.int/ resource/docs/cop1/07a01.pdf.

United Nations Development Programme (2007), *Human Development Report 2007/2008*, New York: United Nations, http://hdr.undp.org/en/media/HDR_20072008_EN_Chapter4.pdf.

United Nations Environment Programme (1990), 'The London Amendment', Nairobi: United Nations Environment Programme, http://ozone.unep.org/Ratification_status/london_ amendment.shtml.

Vanderheiden, Steve (2008), *Atmospheric Justice*, Oxford: Oxford University Press.

Vertovec, Steven, and Robin Cohen (2002), 'Introduction: Conceiving cosmopolitanism', in Steven Vertovec and Robin Cohen (eds), *Conceiving Cosmopolitanism*, Oxford: Oxford University Press, pp. 1–22.

Wapner, Paul, and John Willoughby (2005), 'The irony of environmentalism: The ecological futility but political necessity of lifestyle change', *Ethics and International Affairs*, 19/3 (Fall): 77–89.

WCED (1987), World Commission on Environment and Development, *Our Common Future*, Oxford: Oxford University Press.

Weinstock, Daniel (2007), 'Introduction', in Daniel Weinstock (ed.), *Global Justice, Global Institutions*, Calgary: University of Calgary Press, pp. vii–xxi.

Working Group II of the Intergovernmental Panel on Climate Change (2001), *Climate Change 2001*, http://www.grida.no/CLIMATE/ IPCC_TAR/wg2/010.htm.

WWF (2008), *The Value of Carbon in China*, Hong Kong: WWF Hong Kong/Beijing Office, http://www.wwf.org.hk/eng/pdf/references/press releases_hongkong/WWFcarbon_report_ FINAL_20080630_ENG.pdf.

INDEX

212   WORLD ETHICS AND CLIMATE CHANGE

exploitation, 37, 41, 61, 62, 185
extinction, 21, 23, 25, 69; *see also* biodiversity

fairness, 2, 33, 46, 59, 92, 113, 130, 158, 169; *see also* equity, justice
flooding, 17, 21–4, 26, 38, 110, 114, 135; *see also* climate change
forests, 18, 21, 23, 24, 59, 67, 88, 188
fossil fuel, 18, 19, 36, 59, 89, 110, 141, 143, 169, 172, 173, 175, 179, 186
Framework Convention on Climate Change *see* United Nations Framework Convention on Climate Change
funding climate justice, 169–74; *see also* Adaptation Fund
future generations, 47, 66, 67, 80, 172, 185

Garvey, James, 105, 132, 150, 176, 190
Germany, 94, 129, 137, 142
global citizenship *see* citizenship
global community, 31, 104, 118, 153, 189
Global Environment Facility, 44, 59, 67, 77, 81, 172
global ethics, 2, 35, 36, 106, 184; *see also* cosmopolitanism, justice, world ethics
globalisation, 43, 54, 57, 93, 94, 105, 106, 134, 141, 184
global justice *see* justice
global warming, 3, 17–20, 27, 35, 38, 45, 79, 86–9, 114, 121, 131, 135, 146, 148, 158, 165, 180, 192; *see also* atmosphere, carbon dioxide emissions, climate change, greenhouse gases
greenhouse effect *see* global warming
greenhouse gases, 1, 3, 5–8, 10, 12, 17–20, 23, 27, 34, 36–7, 39, 47, 49, 59, 74–8, 80–2, 84–94, 109, 110, 112–13, 115–16, 118, 121–2, 124, 126, 128, 130–3, 137, 140–2, 144–7, 149–51, 156–7, 159, 161, 165–7, 169–73, 176, 178–80, 182, 186, 188–90, 192; *see also* carbon dioxide emissions

health *see* human health
Held, David, 46, 101
historical responsibility *see* responsibility
human health, 3, 21–2, 23, 24, 35, 40, 46, 47, 61, 140
human rights, 12, 29, 31, 41, 45–7, 50, 101–3, 106, 109, 110, 133–4, 137, 149, 150, 155, 161, 163, 169, 170; *see also* basic rights
human suffering *see* suffering

idealism, 9, 10, 166, 167
ideas, 26, 30, 90, 112, 129, 166, 189
identity, 103, 158, 184
ignorance, 43, 120, 143, 153, 154; *see also* knowledge, science
impartiality, 31, 38, 44–5, 102
individual, the, 2–3, 5–11, 17, 29–31, 35–6, 40–2, 45–7, 53, 55–6, 67, 91–3, 100–3, 105, 109, 111, 113–14, 116, 118–19, 121–54 *passim*, 156–8, 160–1, 163–5, 168, 172–5, 180–1, 183–92; *see also* affluence, capabilities,

consumption, duties, human rights, obligation, responsibility, suffering
industrialisation, 17–19, 22, 39, 49, 59, 60, 65; *see also* Industrial Revolution
industrialised countries *see* developed countries
Industrial Revolution, 18, 144
industry, 18, 21, 34, 91
inequality, 66, 70, 137, 153, 188, 193; *see also* equality
injustice *see* climate change, justice
institutions, 7, 11, 26, 33, 42, 44, 58, 59, 101–9, 111, 115, 118–19, 134, 137, 145–6, 158–62, 168–9, 171, 173, 186, 191
Intergovernmental Panel on Climate Change, 18–22, 24, 26–7, 47, 75, 78, 86–8, 90
international agreements, 5, 8, 12, 36, 55, 67, 75, 93, 163, 167, 178; *see also* international environmental agreements *and specific agreements*
international community, 1, 44, 87, 163
international cooperation *see* cooperation
international doctrine, 4–5, 12, 30–1, 33, 53, 54, 58–73, 94, 99, 100, 108, 114, 149, 152, 157, 158–69, 178, 180, 183, 185, 192; *see also* cosmopolitan corollary
international environmental agreements, 5, 33, 53, 56, 58–75, 78, 182; *see also specific agreements*
international environmental justice *see* justice
international justice *see* justice
international law, 27, 50, 55, 69, 84, 180
international negotiations, 3, 59, 77, 83, 147, 154, 191; *see also* conferences of the parties, diplomacy
international norms, 4, 53, 168; *see also* Westphalian norms
international politics, 8–9, 35, 57; *see also* international relations, world politics
international relations, 12, 30, 44, 56–7, 72, 85, 119, 163, 165
international society, 30, 43, 105, 161, 189; *see also* international community, world society
international system, 1, 2, 54–6, 69, 100, 137, 158, 180
interstate doctrine *see* international doctrine
interstate norms *see* international norms, Westphalian norms
interstate system *see* international system
islands *see* small-island states

Jamieson, Dale, 100, 111, 138, 140
joint implementation, 76, 78
justice, 3, 4, 10–11, 27, 29, 30, 50, 78, 82–5, 90, 92–3, 100, 109, 112, 115–16, 123, 129, 132–5, 151–4, 169, 174, 176, 180
accounts of, 4, 9, 31–2, 37–47, 82, 85
climate, 2, 11, 29, 49, 90–2, 94, 114–17, 120, 122, 125, 128, 130, 132–42, 149–50, 159–60, 163, 168, 170, 175, 178–9
compensatory, 93, 114, 135
cosmopolitan, 2, 100–9, 114, 118–19, 121–2, 133, 142, 154, 156–8, 161, 168, 175
definitions, 4, 32–5